It Started with a Stitch
by Astrid Adams

*For Chris and Brian
Best Wishes
Astrid*

Published by JJ Moffs Independent Book Publisher 2021

Copyright © Astrid Adams 2021

All rights reserved.

No part of this publication may be reproduced, stored in a retrieval system or transmitted in any form or by any means, without the prior permission in writing of the publisher. Neither may it be otherwise circulated in any form of binding or cover other than that in which it is published and without a similar condition including this condition being imposed on the subsequent purchaser.

Astrid Adams has asserted her right under The Copyright, Designs and Patents Act, 1988 to be identified as the author of this work.

JJ Moffs Independent Book Publisher Ltd
Grove House Farm, Grovewood Road,
Misterton, Nottinghamshire DN10 4EF

ISBN 9781838369781

Typeset and cover design by Anna Richards

Contents

Foreword

Chapter 1　The Blyth Tall Ship Williams Gansey Project 1

Chapter 2　Astrid and Janice's Stories 31

Chapter 3　William Smith and his discovery of Antarctica in February 1819..................39

Chapter 4　Blyth Tall Ship Project 50

Chapter 5　Wonderwool Wales..58

Chapter 6　Woodhorn Exhibition 2018...63

Chapter 7　Knitters' Letters and Pictures 69

Chapter 8　Crew to Knitters Letters and Photos...131

Acknowledgements..161

Foreword

When Astrid and Janice wandered into the workshop office at Blyth Tall Ship, little did I know it was the start of something special that would span continents and centuries, inspiring huge numbers of people, drawing on local maritime history and eventually becoming a major part of our organisation's work to transform our community.

That meeting spawned the Blyth Tall Ship Gansey Project. This book is a testament to the commitment of all involved and an inspiration for anyone who might be wondering whether something as unlikely as knitting needles and imagination can change the world for the better.

Enjoy this inspiring tale and may it give you hope.

'An act of love'

I recently had to reflect on the value of our ganseys in the face of a question about their potential commercial market price and came up with 'love'.

It seems to me (given the hours and skill involved) that, historically, few seafarers could have afforded to buy a gansey, despite it being the Rolls Royce of sailing gear. It keeps you warm and the wind and water out like nothing else I've experienced. So, they were knitted to size as an act of love by those who were closest to their seafarer. And they made them pretty and iconic with complex patterns.

And so today the Williams Gansey knitters have transferred that love to the charitable aims of Blyth Tall Ship and its historic celebrations. Every sailor who has received one wears it with pride and appreciation every time they go aboard the *Williams II* for a voyage.

Thank you for your love xxx

Clive Gray
Chief Executive, Blyth Tall Ship

Chapter 1

The Blyth Tall Ship
Williams Gansey Project

In the beginning

In summer 2015 a woman visited the knitting group that meets at Blyth Library every Tuesday morning. The identity of this woman has never been discovered. She asked if any of the knitters would be interested in knitting a gansey for the Blyth Tall Ship (BTS) Project which was planning a trip to Antarctica.

Several of the regulars indicated they would be happy to do so. Janice Snowball attended this group and, always up for a new challenge, she listened with interest. Janice mentioned it to Astrid Adams who, although she did not attend the knitting group at that time, was an old friend and also an experienced knitter. They then heard nothing more.

In August 2016, Blyth was due to host the Tall Ship Festival. Volunteers were being sought to act as stewards for this event. A volunteer event was held in February 2016, which Janice and Astrid attended to see if they could contribute in any way. Clive Gray, CEO of the Blyth Tall Ship Project, was also there with a BTS display. The Blyth Tall Ship Project was to have a key role in the festival, offering support and advice to the ships involved.

Janice and Astrid decided they would ask him what he knew about this request for gansey knitters so, waiting for an opportune moment, they approached him. Clive did not know who had come to the library but told Janice and Astrid he had a vision of having ganseys knitted for the crew of the *Williams II*. This was partly to compare the efficacy of traditional heritage garments to their modern-day equivalent and partly to have a recognisable traditional uniform for crew members.

This was the famous 20 minute conversation during which Janice and Astrid offered to research the history of ganseys, design a bespoke gansey for the crew of the *Williams II*, recruit volunteer knitters to knit the ganseys and raise the necessary funds to do all of this, not knowing that this was going to change their lives. The friends walked away from this conversation wondering what they had let themselves in for.

Although they had both been knitting since they were young and regarded themselves as competent and experienced, neither of them had knitted a gansey before. They had limited experience of knitting design, project management and fundraising.

Of the two Janice was more adventurous in designing her own knitting patterns. Having been a draftsperson in her job, she was well suited to mapping out designs. Astrid had more experience of writing funding bids, articles and project management. That they had such complementary skills was probably the secret of the success of the project.

Having undertaken this task, they first found research material on the internet and by purchasing a number of books. Probably the one most referred to was *Knitting Ganseys* written by Beth Brown-Reinsel; a very comprehensive guide to the history of the gansey, including pattern charts and tips on how to create your own pattern. The Williams Gansey Project drew on this publication extensively.

So, what is a gansey and how does it differ from other knitted jumpers? A traditional gansey is a hardwearing, hand knitted woollen jumper, worn by fishermen around the coast of Britain for many years. The hard twist given to the tightly packed wool fibres in the spinning process and the tightly knitted stitches produce a finish that will repel rain and spray. They are traditionally knitted in one piece with no seams and with diamond shaped underarm gussets to aid freedom of movement; the front being the same as the back. They take between 100 to 200 hours to knit by hand, depending on size, and they can be patterned all over or just the top section and part of the sleeve. The sleeve is knitted from the top down allowing the cuff to be unravelled to repair sleeves that often get the most wear and tear. Usually in navy blue, each gansey had a unique pattern that varied between villages and sometimes between families. Patterns were passed down the

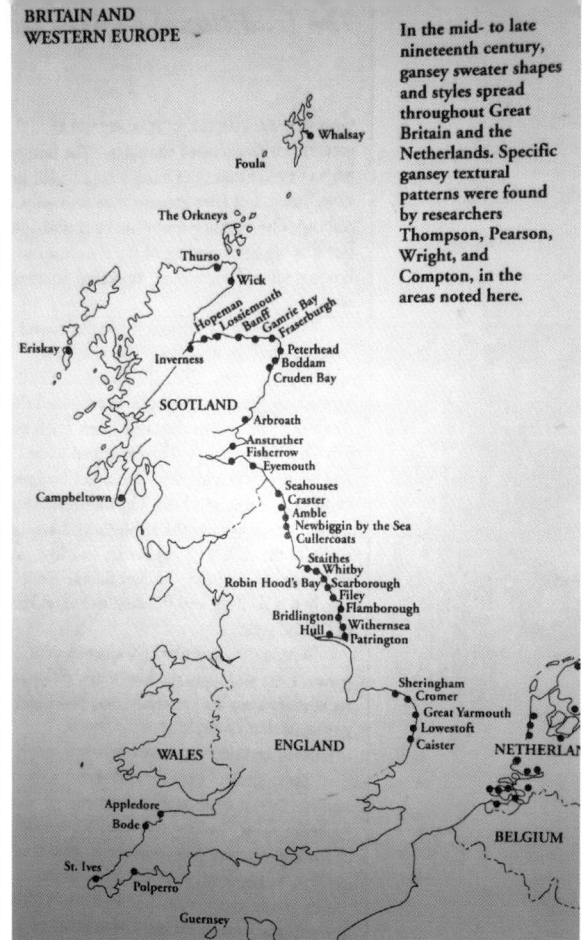

In the mid- to late nineteenth century, gansey sweater shapes and styles spread throughout Great Britain and the Netherlands. Specific gansey textural patterns were found by researchers Thompson, Pearson, Wright, and Compton, in the areas noted here.

generations by word of mouth and were frequently not written down. The individual parts of the design usually represent specific sailing related patterns such as ropes, nets, rigging and waves and can sometimes be figurative including hearts, crosses initials and anchors. They mostly range from the Shetlands, down the East Coast and round to Cornwall but there are some in Ireland, the Hebrides and on the other side of the North Sea.

Although it is impossible to discover precisely when the patterns came into being, they have been traced back to the early 19th century. They were usually knitted by wives and girlfriends, but many were knitted by the fishermen themselves.

The designs

Janice and Astrid's first discovery was that, according to researchers Thompson, Pearson, Wright and Compton, Blyth never had its own gansey design (see map). Janice and Astrid felt this gave them a blank sheet to start with. They agreed that the gansey should be knitted in the traditional way, all in one piece with underarm gussets and no seams.

The pair agreed to combine traditional designs with more innovative ones. They decided they would like to have the logo of the BTS Project included in the overall design and also felt that it would be appropriate to include the Northumberland flag. They believe this is the first gansey to include a logo or a flag, thus evolving the tradition of gansey design. Janice designed a pattern that she called the 'Blyth Staithes' pattern to represent the wooden structures that used to run alongside the River Blyth carrying trains with coal for export. The *Williams II* is moored alongside the remains of the staithes. The ganseys also had the initials of the Blyth Tall Ship Project just above the welt. The photos show the designs on the grey gansey as the details are clearer on these.

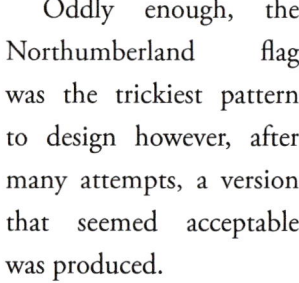

Janice and Astrid spent the early months of 2016 working on the patterns and how they would fit into the finished gansey.

The logo became a feature of the gansey. It was soon clear that the pattern was so complex that it would have to be quite large and that the best place for it on the gansey was a central panel.

Oddly enough, the Northumberland flag was the trickiest pattern to design however, after many attempts, a version that seemed acceptable was produced.

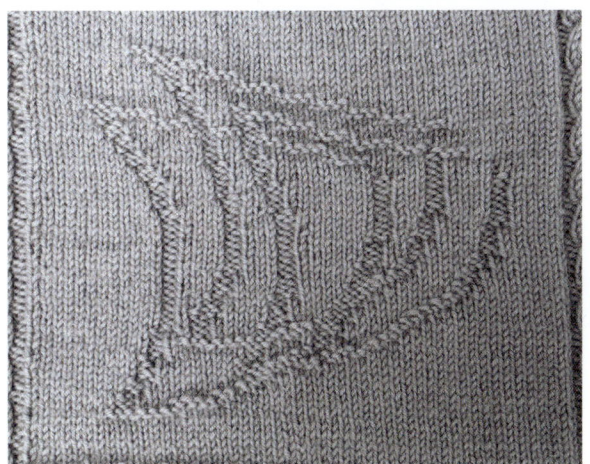

It Started with a Stitch

The pair had regular discussions with Clive Gray, informing him of progress, showing samples and discussing possible funding. The Friends of Blyth Tall Ship (FOBTS) group also responded very positively to the progress of the project.

Advice was taken on where to apply for funding but initially this was unsuccessful.

A friend (Sue Andrew) managed to raise £200 from a local Masons' group. Janice and Astrid were very grateful for this as it gave them the means to buy the first four cones of five-ply, navy-blue gansey wool from Frangipani, based in Cornwall, who went on to supply all the wool for the project.

The first design was laid out on an Excel spreadsheet with a square per stitch. When it was first printed out, it covered nine A4 sheets that were taped together and had to be spread out on the floor to read. Janice produced a drawing of what the finished gansey should look like.

Sue Andrew

Labelled gansey impression

Janice and Astrid began knitting prototypes, never giving up on raising the necessary funding, despite early disappointments. They bought two sets of five steel pins, intending to knit in the traditional way, however Janice described knitting with these needles as "trying to wrestle with an octopus". They decided circular needles were a safer option.

Funding

Janice and Astrid had a passing acquaintance with Grant Davey the (then) leader of Northumberland County Council (NCC) who agreed to meet them to discuss a Community Chest Grant for the project.

A meeting date was fixed for 13 October 2016. As the meeting date approached, Janice and Astrid worked hard to knit as much of their prototype ganseys as they could and to develop a project plan so they could demonstrate that their idea was sound and the pattern would meet the needs of the Blyth Tall Ship Project. The grant application included projected costs, including printing, materials, postage and packing, circular knitting needles and labels. They were asked to include volunteer

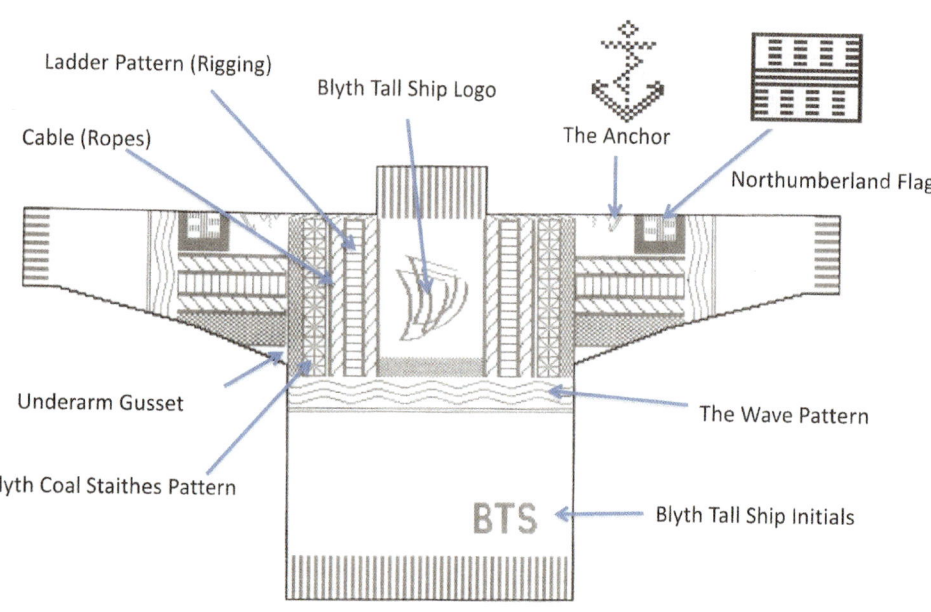

hours projecting the time it would take to knit each gansey at an average of 150 hours to demonstrate the contribution to the project that would be made by the volunteer knitters.

Towards the end of November 2016 they were informed their bid had been successful and that of the £8,000 they had requested, they were given £7,771. This was such exciting news that the pair did not delay in sharing it with BTS CEO Clive Gray.

The next step was to begin recruiting volunteers. Janice and Astrid had been resigned to undertaking a series of visits to as many local knitting groups as possible to try to find people who would agree to knit a gansey for the Williams Gansey Project.

Northumberland County Council gave the pair extensive help and support with the next phase. They agreed to print patterns and grids for them at no cost to the project, which saved a substantial amount of money and made it possible to knit more ganseys with the money available.

They also helped with press releases and publicity, as did Astrid's sister, Claire. The details of the project were published on the NCC web site and on social media platforms. The result of this was astonishing. Within two weeks Janice and Astrid received almost 600 contacts either by email or Facebook from knitters wanting to be involved in the project.

The pair realised that rather than recruiting knitters, they now had to shortlist from this huge list. They took on this task separately, each taking a copy of the list and working through the applications to knit – awarding one, two or three ticks to each one, depending on the content.

The applications came not only from UK knitters but also from knitters all over the world. It had never been intended to go outside the UK, however the response to the publicity led them to agree a limited number of knitters should be included from every continent possible. In the end only the continent of Antarctica was not represented.

It was agreed that priority should be given to knitters from Blyth or those having links with Blyth, then moving out to Northumberland and the wider North East. They wanted to have as many areas of the UK represented as possible. They also included knitters with links to the sea, to Antarctica and with family links to historic exploration.

Happily enough, when Janice and Astrid got back together to review their choices, they found they had largely agreed on their selections, never being more than one tick apart.

7

Knitters' applications

Below are extracts from just some of the applications to give an idea of what the pair were looking for and the range of applications received.

✓ *"I have been knitting for many years and have some experience knitting on circular needles though I must admit that I prefer to tuck one needle under my arm! Here's to a very successful project. I look forward to following its progress as my husband's grandfather and great grandfather were both ships' captains, the former having sailed from Blyth on some occasions."*

✓ *"I am an experienced knitter, having knit ganseys (and hats, socks, etc.) for my late husband. I am a sailor, both of a yacht and a canoe! I am fascinated by early exploration of the Arctic and Antarctic, the more so having just got back from Arctic Norway. My late husband would have been as excited as I am to learn about this trip. I would love to volunteer to knit a gansey and a hat. I can think of no better way of combining my interests to help this trip on its way and for me, it would also be a way of remembering my late husband."*

✓ From a male knitter – *"A friend of mine sent me the article about the gansey project. I have just knit a gansey, and am knitting my wife another jumper. Would love to knit one of your ganseys. I started knitting last November. My wife who's an excellent knitter started me, but it was my mother who taught me to knit as a child nearly 50 years ago. She knit a few gurnseys herself. She died last July and I often think of her while knitting."*

✓ *"I live in Bridgnorth, Shropshire. My wife is from Cramlington. We lived there for five years then moved to Berwick and were there for seven years. Work brought us down to Shropshire 1998. We come back to Northumberland often as we have family and friends there and just because we love it."*

✓ *"...yesterday was the 50th anniversary of my joining the WRNS so, although we never went to sea, it's definitely in my heart. Add to that, being married to a sailor who loved tall ships and being a very happy knitter, I hope to hear that you still need volunteers and don't mind where they live!"*

✓ *"I would love to have a go, but live in New Zealand. I have been knitting for over 60 years and have made lots of jerseys with Aran patterns - have never heard of ganseys before but sounds interesting."*

✓ *"I am a knitwear masters student at Heriot Watt University in Galashiels and found out about your wonderful campaign for handmade fisherman's ganseys to go to Antarctica! I did a small design project influenced by Scott's trials in Antarctica in 1912 and am an avid handknitter. This project seems right up my alley!"*

✓ *"I'm interested for two reasons. 1) I'm a keen knitter and have been knitting since I was 18 years old (both self-taught and 30 odd years' experience) 2) I have a son who is a keen sailor. He sailed around the Arctic Circle for his Duke of Edinburgh gold award. He has gone on to achieve a lot of different qualifications as well as many experiences, one of the first being a volunteer with the Tall Ships and Ocean Youth Trust and has had experience sailing/working on tall ships.*

✓ *"I am also a keen enthusiast on everything to do with Antarctica, having read books and journals of Amundsen, Shackleton and Scott. As a proud Canadian, I have been interested in the Franklin Expedition to the Arctic in the 1840s, as well as the recent discoveries of the final resting place of both men and ships".*

✓ *"I hail from Halifax, Nova Scotia, home to Bluenose and Bluenose II, so you could say that the sailing life is in my blood and I have attended the Tall Ships gatherings in the past. I now live in Calgary, Alberta with my family and two beautiful cats. I would be honoured to knit a sweater or two for the brave members of this expedition and I can begin knitting right away as I am retired and have time to dedicate to this project."*

✓ *"I am from the Outer Hebrides (The Western Isles of Scotland) and I am a knitter of our own version of the fisherman's gansey, known as the Eriskay Gansey after the tiny island of Eriskay which has kept the tradition alive. I live on the neighbouring island of South Uist and was taught how to knit ganseys by two women from Eriskay. I have since made seven jerseys and I like to have one on the pins in case I forget how to do them!"*

✓ *"I have been knitting for nigh on 60 years and it is one of my major hobbies. Every morning before I get dressed I do at least one hour of knitting."*

✓ *"I am a graduate of the Royal College of Art London with an MA in knitted textiles. I have just seen your Facebook post and would love to be involved in the project."*

These and other stories, were to have an unexpected and profound impact on Janice and Astrid and influenced their future actions. The project was giving an opportunity to volunteer knitters to be involved in the wider BTS Project in a way that would otherwise have been denied to them. It was agreed knitters would be asked to send photographs of themselves with their completed ganseys and crew members, once the ganseys and hats had been allocated, would be asked to write to their knitter, via the project, to complete the circle

The launch

By December of 2016 the pair had completed the first two 'prototype' ganseys, which they took to the official launch of the Williams Gansey Project on 7 December 2016 at the Blyth Tall Ship Christmas party. Clive Gray tried on both of the ganseys and decided that Astrid's fitted him best. He asked if he could keep it to wear at talks and presentations. Astrid said later that it almost felt like a sort of bereavement leaving her precious gansey behind. She had to go home and begin a second one immediately.

Gansey knitters will understand that feeling. The work that goes into knitting a gansey, which may take up to 200 hours, makes it a very personal experience.

Getting the grant money meant that Janice and Astrid were now able to buy the materials needed to put kits together to send to the selected knitters.

At the same time Janice and Astrid had been working to transfer the gansey pattern from an Excel spreadsheet to a five-size gansey knitting pattern. This was no mean task. The gansey pattern is fairly complicated and the pair had no previous experience of this. However, Astrid began the task, sending it as a work in progress to Janice at intervals for checking. They found that, no matter how frequently they checked the pattern, they would still find mistakes.

Once they had a first draft, Janice, Astrid and Doreen (a lady from the Tuesday knitting group) began the task of knitting ganseys in the five different sizes, amending the pattern as necessary as they went.

Eventually it was felt the pattern was correct. Despite this a few minor errors were still found later on. Some parts of the pattern were particularly difficult to translate into the written pattern (especially in five sizes) so parts of the pattern (including the BTS logo and the anchor) were sent using an Excel grid in just one size. This meant that, as patterns were sent out to volunteer knitters, they would get a pattern in five sizes but, for part of their knitting they had to refer to the grid for the size they were being asked to knit.

Male gansey knitters

It was with surprise and delight that Janice and Astrid received offers to knit from some male knitters. In the past sailors often knitted to relieve the boredom of long sea voyages but, while it may sound sexist to say, it is not often one sees men knitting (although Astrid's son Joe does some wonderful crochet work). The first two offers to knit a gansey from men came from Philip Batten and Davie Jeffrey. The third was from Richard Sloggett who approached the pair at Wonderwool Wales with an offer to knit. He (and his family) returned the following year to drop off the completed gansey. Perhaps seeing Tom Daley knitting at the 2020 Olympics (held in 2021) will encourage more men to pick up the needles and knit.

Gansey central

A room in Astrid's house became "gansey central". The biggest delivery was eight huge boxes of gansey wool. Putting all the kits together was a bit of a logistical nightmare.

The wool arrived in 500g cones and, as the smaller sizes did not take two full cones, Astrid set to the task of winding wool so the right amount was provided for each size.

Each kit had to be checked to make sure everything was included. Some knitters wanted circular needles included and some did not. There were two different sizes of needles depending on the tension of individual knitters and everything had to be logged on a database.

For kits coming back from UK addresses return postage was included. This was not possible for those coming from abroad as postage costs were not known, but these were reimbursed if requested.

Post Office staff in Blyth were surprised when, in May 2017, Janice and Astrid arrived with several bags full of gansey kits to post and did suggest they come at a less busy time in future. However, they were generally tolerant of the quantity of post and didn't quite run and hide when the pair came through the door.

The initial aims of the Blyth Tall Ship Project have had to change over time. At the beginning of the project the intention was to recreate the voyage of William Smith in February 2019, the bicentenary of his first discovery of Antarctica (see Chapter 3).

There is no doubt the idea of being involved in a project that would provide ganseys for crews sailing on the four legs to Antarctica had caught the imagination of the prospective knitters, as indeed it had done with Janice and Astrid when they first set up the project.

However, even if the

voyage had gone ahead as planned, only a very few ganseys would actually have reached Antarctica; the number needed would have been no more than around 80. Volunteers were informed that at best, only around 20 ganseys would be on the final leg to Antarctica itself. Because of this, knitters were given the option of withdrawing their offer. Thankfully no one did, all still wanting to remain involved in the project.

There was to be a training voyage round Britain early in 2019 to ensure crew members sailing to Antarctica were fully prepared for what would be a very testing expedition. Because of the level of response, Janice and Astrid agreed to ask knitters if they would be prepared to knit ganseys for the Round Britain Trials instead of specifically for the Antarctic expedition. Without exception, all those approached agreed they would be happy to do this. This allowed Janice and Astrid to increase the number of knitters from 80 to 125.

The difficulty they faced was they had no idea who the crew for the Round Britain Trials would be. They had to hope that knitting in a range of sizes, with fewer in the largest and smallest sizes would be sufficient to fit out the 10 crews as they were recruited.

The Williams gansey watch hat

Jackie, one of the knitters at the Tuesday knitting group, said that she would never be able to knit a gansey, but she might manage a hat. She was not alone. Quite a few knitters emailed suggesting a hat would be more manageable for them. With this in mind Janice set about designing the Williams gansey watch hat. This increased the number of people that could be invited to knit and also meant that all those who sailed on the Round Britain Trials could be provided with a hat as well as a gansey.

The hat includes many of the patterns from the gansey, such as the Northumberland flag, the cable and the ladders. There were male hat knitters too.

To include the knitters more fully in the project and to link them to the crew member who received their gansey or hat, Astrid designed a label to be sewn into the garments on which the name of the knitter and the crew member receiving the gansey could be included. This also helped on board the *Williams II*, because there was always the danger of a mix up with so many identical ganseys in a confined space.

There was an expectation that the crew member would contact their knitter to let them know about their experiences on the *Williams II* and why they had applied to be involved in the project.

It Started with a Stitch

Joe Dowling

Russell Pettit

Vera Smith

Gansey workshops

In April 2017, Janice and Astrid arranged a distribution day, where local knitters could come and pick up their gansey kits, meet other knitters and see the completed prototype ganseys.

In August 2017 a workshop for local knitters was held so that knitters could meet each other and discuss any difficulties they were experiencing. One area of the gansey that seemed to cause problems was the shoulder panel (a band of knitting that joined the front and back of the gansey). After hearing that knitters were finding instructions for the panel difficult to follow Janice re-wrote it in a simpler form.

Blyth Library

Blyth Library very kindly agreed to receive returned parcels posted within the UK. This was extremely helpful as it meant there would always be someone 'at home' when the postman delivered. It also led to a sense of excited anticipation at each Tuesday morning knitting group to see if any parcels had been delivered. Letters enclosed within the parcels were read out to the general delight of the wider group. The staff at Blyth Library have been very supportive of the project, inviting Janice and Astrid to put up a display in May 2018 followed by a talk in July 2018.

Completed ganseys returning

The first completed UK gansey was collected on 8 June 2017 and was knitted by Kay Atkinson. The first completed gansey from abroad was returned by Sandra Marie Boock from New Zealand and was received in November 2017.

Emails were sent to the knitters to acknowledge the receipt of the gansey and to thank them for being part of the project. The ganseys were logged, measured, bagged and put in boxes according to size.

Some knitters took a year or more to knit their gansey and some were returned unfinished for various reasons. The navy blue wool proved too difficult a colour to knit for some and other knitters experienced personal difficulties and could not complete their ganseys. A few wonderful local knitters were recruited to complete these ganseys. The names of both knitters were included on the label. Janice and Astrid were determined that being involved in the project should not in any way become a chore or cause the knitters any distress, so were quite comfortable with these returns.

By the end of August 2017, the hat pattern was finished and had been tested. Janice and Astrid set about putting together hat kits to post out to volunteer knitters.

At this time Janice and Astrid were designing a display about the gansey project for an exhibition at Woodhorn Colliery Museum near Ashington in Northumberland that was being held from February to May 2018 (see chapter 6).

Talks and visits

The pair have given a number of talks to local organisations. These have covered a wide range of groups from knitting circles, Mothers' Unions, Women's Institutes, community groups and a talk at Blyth Library. One of the most challenging was when they were asked to talk to a group of five to seven year olds at a school near Hexham, Northumberland. It was difficult to engage a fairly large group of young

children in a talk about knitting, although the teachers were very supportive. They were invited twice to talk to users at the Northumberland Blind Association. Janice and Astrid wondered how this would go as their talk about the gansey was very visual. However, ganseys were passed round the group and patterns were felt by the fingers of those who were unable to see them. Fortunately the gansey is very tactile and group members appreciated the talks.

From March 2018 they also arranged for several groups of knitters and their families to visit the BTS Project when they were able to get onto the *Williams II*.

The gansey flag

Janice also designed a 'gansey flag' for an event organised by Headway Arts based in Blyth; an organisation that sets out to create a cultural hub, providing opportunities for participation in various arts that can be enjoyed by the local community. They asked community groups to design their own flag for display on Blyth beach in August 2018 and, never afraid of a challenge, Janice came up with this design.

Planning for the Round Britain Trials

In January 2019, it was eventually decided by the BTS Board that the voyage to Antarctica in the *Williams II* would not go ahead in 2019. Although this was very disappointing, there were several reasons for this decision. Not least of these was because of the impact of the impending Brexit that was making businesses reluctant to invest in UK projects. There were also concerns about the safety of taking a 100 year old ship on what might be a perilous voyage. The *Williams II* was still undergoing refurbishment but was by no means sufficiently seaworthy for this sort of test. The Round Britain Trials were still to go ahead from March to May 2019. There would be 10 legs to this voyage with crews changing every Saturday. The logistics of this were complicated. As crew members were appointed, Astrid contacted them to ask for measurements so they could be matched to a stored gansey and watch hat.

Third anniversary

In February 2019 Janice and Astrid marked the third anniversary of their first conversation with Clive Gray with a celebratory meal together.

The Williams Gansey Project had grown bigger than either of them could have expected. The Round Britain Trials were to take place soon and there were sufficient ganseys and watch hats in store to accommodate all crew members. The only concern was that there would be enough variation in sizes to fit the 10 crews.

Materialistics

A group of knitters called "Materialistics" based at the library in South Shields offered to knit ganseys for the project. In June 2019 Janice and Astrid travelled to South Shields to collect the completed ganseys and give a talk about the project.

Gansey distribution

As many ganseys as possible were tried on and given out before the *Williams II* set out on the Round Britain Trials. The range of sizes knitted enabled all shapes and sizes of crew members to be kitted out - everything from chest size 34" to 54".

Matching prospective crew members with a

gansey and watch hat was a time consuming and difficult task. The *Williams II* is not a large ship and it was difficult to find a place to store the ganseys

and hats for the joining crews. Eventually it was agreed they should go in a storage cupboard in the captain's cabin. Janice and Astrid spent the best part of three days leading up to the Round Britain Trials sorting ganseys into boxes to go on board the *Williams II* for each leg. Each box had a list of the

crew members to whom the gansey and hat were to be given. The individual ganseys and hats were in plastic bags. They contained a letter for the crew member informing them of washing instructions, telling them who their knitters were and asking them to take the time to contact their knitter (via the project) to let them know who they were, why they had become involved and what it had meant to them to be part of the crew of the *Williams II* on the Round Britain Trials. It was little short of a miracle that, of the 125 or so ganseys and hats knitted, Janice and Astrid were able to match everyone with a gansey and hat to fit. They suffered some anxiety about whether the distribution of the ganseys and watch hats on each leg would be carried out without losing any of them. However, despite the difficulty of the task, only two ganseys appear to have gone missing during this process.

The grey ganseys

There are a number of people working as volunteers at the Blyth Tall Ship Project. Many are retired and their input into the project is invaluable. They pass on their skills to trainees, generously giving their time and experience to the project.

It seemed unfair that this group would not be given a gansey, but it had been agreed the navy blue ganseys should only be given to those sailing on the *Williams II*. A decision was made to try to raise further funds to provide grey ganseys for this group. To this end Astrid began researching crowdfunding. She put together a short film about the gansey project and the aim of knitting grey ganseys and hats for this group of volunteers, as well as the small group of people who were involved in recruiting the crews for the Round Britain Trials. The crowdfunding appeal was launched in July 2018. The generosity of supporters of the Williams Gansey Project led to the raising of £2,500. This was enough to buy the materials necessary for the grey ganseys.

Knitters were invited to knit the grey ganseys and once again the pair were delighted with the response. They were looking to knit around 30 grey ganseys and watch hats. These kits were posted in October and November 2018.

By May 2019 Janice and Astrid were ready to begin handing out some of the grey ganseys to volunteers in the workshop. Grey ganseys were also received by Janice and Astrid (although Janice knitted her own).

Martin Haigh

Kevin Twaddle

Kate Mojaat

Astrid's thankyou letter to her knitter Doreen Armstrong

Hi Doreen - as promised I'm writing to let you know that I have the grey gansey that you knitted for the project.

I think you know what the project has meant to me and Janice. One of the high spots so far was being invited onto the ship for its return to Blyth on 1 June. The icing on the cake was having my own gansey to wear.

I had always intended to knit myself a grey gansey but because I'd already finished off two navy ones this year, I could not bring myself to pick up the needles to start another one.

The whole project has been a roller coaster of a journey for me and Janice. As you know, we never expected or intended it to be so wide reaching. The support we have received from people like you has made the whole thing not only possible, but so rewarding.

The knitting group at Blyth library has been central and very special to the project as this seems to be where it all began. It is a little oasis of calm in the middle of the week.

Once again thank you for your contribution to the project, which you have supported right from the beginning.

Best wishes

Astrid

The Round Britain Trials

On 17 March 2019 an open day was held at the BTS workshop. Local TV and media had been invited. This was the planned departure date for the *Williams II* on the first leg of the Round Britain Trials. However, the departure was delayed by high winds that made it unsafe for the *Williams II* to leave the river. The ship eventually sailed at 5.30am on 19 March, watched by a small group of supporters. Astrid kept the Williams Gansey Facebook page updated, regularly reporting on the progress of the ship using a ship tracking app. Photographs were also included where they were available.

One of the first photographs was of the ship's cat, Cuthbert (a glove puppet) the only crew member to be aboard for the entire voyage.

It immediately struck Janice and Astrid that Cuthbert did not have a gansey and hat. Janice set about resolving this situation, spending the following few days knitting these garments for Cuthbert. Janice even got up at 4.30am to pass these items over to the crew leaving Blyth the following Saturday to sail on the second leg from Lowestoft.

The photographs from the following week showed that Janice had knitted the right size for Cuthbert. To the delight of the pair Fiona Williams, one of the hat knitters, sent a couple of knitted mice

It Started with a Stitch

for Cuthbert to play with.

When the ship reached Holyhead, Anglesey in April, Christine Jukes, one of the knitters, decided she would travel down to where the ship was moored and introduce herself.

Christine was thrilled to be invited to sail with the crew for the day. She and a friend, who had gone with her, spent several hours sailing on the Irish Sea. This was over the Easter weekend and, as there were difficulties in getting supplies delivered to the ship, Christine kindly drove some of the crew to the supermarket and back with supplies.

A number of group photographs of the crews were taken, showing them proudly wearing their ganseys and hats. Crew members have said that, when they wore them in port, they received much

admiration during crew outings on shore.

In what was a hugely successful journey, the *Williams II* covered 1,930 miles, calling into 25 harbours and anchorages around Britain, including three countries (or four, if you count the Isle of Mann as a different country). Over the 10 legs, 130 berths were filled and over 30 of these were taken up by young apprentices or crew members who were unemployed or in some way disadvantaged. At least 10 young people completed what would eventually be their RYA (Royal Yachting Association) Competent Crew certificate. The maximum speed through the water was about 7.5 knots as the *Williams II* sped home down the Northumberland coast.

The *Williams II* actually returned to the river a week before the welcome back open day. Janice and Astrid were delighted to be invited to be on board the ship when she officially returned on 1 June 2019. Wearing their grey ganseys, they sailed out into the North Sea during the morning and returned to the River Blyth with a flotilla of yachts, boats and RNLI lifeboats. All of the crew were wearing their ganseys, making it one of the proudest moments of Janice and Astrid's lives. Sailing on the *Williams II* for such a significant event was a fitting end to the first phase of the Williams Gansey Project. Janice and Astrid (with the support of their amazing knitting community) had achieved what they set out to achieve. Each crew member taking part in the Round Britain Trials was supplied with a Williams gansey and a watch hat.

It Started with a Stitch

Rewriting the knitting pattern

Janice and Astrid had always said that, until the completion of the Round Britain Trials, the Williams gansey pattern would not be for sale. They felt (as did Clive Gray) that the ganseys should be solely for the crew members of the *Williams II* until that voyage was over.

So it was in June 2019 that Janice began the task of preparing the gansey pattern for sale. The main differences from the earlier pattern was that every part of the pattern had to be written in five sizes and, although grids were available with the pattern, written instructions were to be included as well. This was by no means an easy task, but Janice set about it with her usual methodical determination.

Between them Janice and Astrid have also designed a scarf, fingerless gloves and a sock pattern in the Williams gansey style. Patterns are available for all of these (see end notes). There is also a cushion pattern designed by one of the project's wonderful knitters, Iris Scott. She has also designed a Williams gansey dog coat for a small dog.

Countryfile

In August 2019, the BTS Project had a visit from the BBC Countryfile team. Two years previously Astrid had seen an item on Countryfile, introduced by John Craven. He was interviewing a lady who had knitted a gansey for her husband every year of their 25-year marriage. During this interview he had described gansey knitting as a "dying art." Having received so many offers to knit a gansey from knitters all over the world, Astrid felt she could not let this go unchallenged. She emailed the Countryfile team telling them about the BTS Project and how the Williams Gansey Project had come out of this. Around 18 months later, when Astrid had given up on hearing anything back, she received an email from the programme asking if they could come to film. The filming was delayed a couple of times and there were times when it appeared it would not go ahead at all. Eventually in August 2019, the film crew with Helen Skelton visited and spent all day filming the various aspects of the project. Shown on 9 September 2019, the BTS and gansey projects were given the first and last five minutes of the show, with the gansey project coming last, but by no means least.

The Knitter magazine

In January 2020 the gansey project was featured in the Knitter magazine. Astrid had been approached by Penny Batchelor who wrote an article based on communications with Astrid and her own research into the project.

The ganseys that reached Antarctica

In February 2020 Liz King, who had been skipper on the *Williams II* for eight of the 10 legs on the Round Britain Trials went to Antarctica on the *Europa*. She had her gansey with her and she also took Cuthbert, the ship's cat (with his gansey naturally) so two Williams ganseys did make it to Antarctica after all.

Liz King writes "Being Skipper of *Williams II* during the Round Britain Expedition in spring 2019 was a hugely challenging and rewarding experience. From March to May 2019, we covered almost 2,000 nautical miles. Giving out the ganseys to the joining crew members was one of the highlights of each leg of the Round Britain journey. I am immensely fond of my gansey and nowadays I wear it with pride on every boat I sail. In particular, I was very proud to wear the Blyth gansey during my trip to Antarctica in February 2020. My vessel for my Antarctic trip was the bark "*Europa*." Originally built in 1910 and used as a lightship protecting the estuary of the river Elbe, *Europa* is a 56m steel bark, carrying 27 sails as well as two huge Caterpillar engines. Along with the *Williams II* mascot "Cuthbert the Cat" I boarded *Europa* in Ushuaia, Argentina, on 1 February 2020, accompanied by 45 fellow passengers for the three-week trip to Antarctica. Sighting what are now known as the South Shetland Islands on 6

February 2020 brought an emotional lump to my throat. I could only imagine the astonishment and excitement on board the original Williams, 200 years previously, when a similar sight was first seen by the lookouts. This was the first land to be sighted beyond 60 degrees south. Not even James Cook, 40 years previously, had seen land that far south. It was with huge pride that Cuthbert and I hopped ashore on the South Shetland Islands – and eventually on the Antarctic continent – wearing our Blyth ganseys."

post or electronically in PDF form.

At the beginning of this project in 2016, Janice and Astrid would never have believed what the future would hold. The support and encouragement they have received from their knitters and the wider community has been astonishing.

The Williams Gansey will remain the bespoke uniform for the crews of the *Williams II* on any future major voyages and is a testament to the dedication of the knitters who volunteered their time and skill to be part of the Williams Gansey Project.

A display of the Blyth Tall Ship Gansey Project and an exhibition of photographs and other information from the Round Britain Trials can be seen at the Harbour Commissioners building, Bridge Street, Blyth.

This brings us up to date to the time of writing this book. The Covid 19 pandemic in 2020 brought everything to a full stop, although knitting patterns have been dispatched at a fairly steady rate, either by

Europa in the ice

Chapter 2

Astrid and Janice's Stories

Astrid's story

I was born in July 1950, the third of what would be six girls. I think my poor father believed that if he kept trying he might have a son.

We didn't have a lot of money when I was growing up and I remember well my mother buying "cut-out-and-ready-to-sew" dresses for us from women's magazines. I guess that's how I learnt to sew from her, though I can't really remember ever doing so.

I was educated long enough ago to have been taught how to knit at school, although I was not a keen knitter until much later in life.

There were two clear strands to my upbringing. My father, a civil servant, was also a lay reader in the Church of England, so Sundays were always church days and in turn I think I and most of my sisters eventually joined the church choir. Music (singing) has played a significant part in my life. As children, we spent happy times singing round the piano, my father playing. His musical taste was eclectic, from choral to jazz. Although not quite the Von-Trapps, as a teenager I sang with some of my sisters in an acapella folk group. We still do sing when we get together – it's in the blood.

The other strand was political. My father was a member of the local Labour Party and I spent many hours folding election leaflets and putting them in envelopes ready for posting. I think this grounding gave me a strong social conscience and made me into the sort of person who gets involved, not a person who sits on the sidelines letting others do the work. This resulted in many things including several visits to the American cruise missile base at Greenham Common for planned protest events.

I am also a 'crafter', and have dabbled with dressmaking, machine and hand sewn quilts, card-making, cake decorating, crochet and most recently cross stitch. These are all self-taught and I would describe myself as a jack-of-all-trades.

I was educated at a time before the comprehensive education system came into being. I passed the eleven-plus and began my grammar school education at Gosforth Grammar School. However, in the third year (1963), when I was 13 years old, my father was offered an experimental role as a Church of England lay reader to manage the parish of Ellington in Northumberland. The parish church was at Cresswell. We were living in a huge rambling vicarage set in its own grounds with attic rooms and a cellar. One of the rooms had been turned into a small chapel. Sunday services were held there three weeks out of four and a bus was hired to take the congregation to the church at Cresswell on the fourth week. There was also a parish room at the back where village events took place – including a local youth club. This made us very popular with young people locally. The front lawn was large enough to hold the village fete and there were stables and a pigsty to the rear. A small orchard was located to the side of the site.

When we moved in 1963, I went to Ashington Grammar School, which was very different from Gosforth. Coal mining was still the main industry and employer. It was during this time at Ashington Grammar that I met Janice. You can probably imagine the sort of high jinx we used to get up to in the vicarage, but this is probably better glossed over.

We were only in Ellington for three years and moved back to Gosforth in 1966. Janice and I kept in touch until we were both married and had started our families, but lost touch over the years.

I was living with my (first) husband in a council flat in Gosforth at the beginning of my marriage and we were soon offered a council house in a new estate. When thinking back, this is probably the beginning of my "serial volunteering" (as Janice puts it). There were some difficulties with the houses on this new estate. They had flat roofs and were built of light yellow bricks. The cement that had been used turned out to have too much sand in it and there were problems with damp coming through. The flat roofs were causing a similar problem. It was agreed we needed a tenants' association to make representations to the council. I became secretary of this association. I seem to have become secretary of just about every group I have joined ever since, most recently being secretary of the Friends of Blyth Tall Ship.

I moved to Whitley Bay in 1979, buying my first house with my first husband and my mother, who lived in a 'granny' flat at the top of the three-storey house.

The house was always full of children. I had three

of my own at the time and I fostered as well. My many nieces and nephews were also frequent visitors. I had a fourth child later on, so I had two boys and two girls. I now also have five grandchildren and five great grandchildren.

I became involved in a local church, being one of the Sunday school leaders for a time and of course secretary of the PCC (parochial church council). I was a key contributor to a small group that raised half a million pounds to build a new parish centre attached to the church. I was also a governor at the middle school that my children attended.

Skipping through the rest of my time in Whitley Bay, including two more house moves and a second marriage, in 2010 (having just got divorced for a second time) I moved to Blyth. I got back in touch with Janice who had never moved from the address I last had for her and like a blast from the past, came back into her life.

As she mentions in her story, it was a bit spooky how many similarities there were in our lives, our hobbies and our children's interests. Despite the many years we'd not been in touch, we soon made up for lost time, getting involved in the local theatre as volunteers. Initially we both worked with wardrobe, but I moved onto props and she concentrated mainly on making bits and pieces, fairy wings being a speciality. Like Janice, I am also a member of Friends of Ridley Park, providing support and helping with the planning of events in our local park (my sister Claire is secretary of this group). The rest, as they say, is history.

We may never know the identity of the woman who came to the Blyth Library knitting group in 2015, but we have a lot to thank her for. The Williams Gansey Project has become such a key part of our lives and I think it's probably true that there have been no other gansey-knitting projects on this scale. Although the initial aims of the Blyth Tall Ship Project have had to be revised and altered over time, the Williams Gansey Project has achieved everything it set out to achieve. Janice and I have both been astounded at the level of support we have received from our knitting community and are so very grateful for the contributions made by all our knitters.

Janice's Story

I'd like to start by saying the Williams Gansey Project was not my idea. I just helped to make it happen. I'm not a go-getter, not the sort of person who goes looking for a crusade or even just something to do. Things just sort of happen. I respond to circumstances and situations, though lately I've taken to describing myself as 'a serial volunteer'. It's not my fault. I blame my parents. Let me explain.

When I was only two weeks old, we moved into a new house on a new estate in Newbiggin by the Sea in Northumberland. At the bottom of our garden was a patch of spare land and within a few years the council had built a hall on it. This was used mainly by a Senior Citizens' Club. Soon my parents started attending their Saturday coffee mornings, often taking along a baked contribution. When they had a fair to raise funds, Mam made a couple of cakes to sell and she and Dad started making a few other things such as aprons (Dad was a dab hand with the sewing machine as well as Mam). The Senior Citizens' Club held fairs twice a year and it wasn't long before my parents had their own stall selling aprons, sponge bags, novelty soap holders and 'dressing table' dolls. Almost everything sold. I was the youngest of four girls and we all helped with the making. We weren't forced to; it was a joy to make these things with them. So that's how it all began. Dad also painted the interior of the hall and at Christmas we made and strung up decorations.

Then my next youngest sister and I started dancing classes. We did many shows, mostly for aged peoples' clubs, homes for the blind, the disabled etc. The costumes were all home made by the dance teacher and her mother. Mam and Dad started helping in a small way with these – mostly with accessories, though they helped in a big way when it came to decorating a float for the local carnival parade. To the family's surprise they even got involved on stage in an adult singing group. All of this wasn't high class stuff, but I realised then just how much it meant to the people to whom we were performing. Our payment was a cup of tea and a biscuit, though some clubs put on a good spread. We often had a chat to our audiences afterwards too, which made us feel quite special. The seeds of the rewards of doing something for the benefit of others were really starting to grow!

Things quietened down when I got married and left home. It wasn't until I had children of my own and they started school that it all began again. I was a parent helper on many school visits or events in school. When my son, the eldest, started high school he was invited to join the newly-formed Young

Engineers' Club. They were going to take part in a competition in London but, as there were girls in the club, they needed an adult female with them and no members of staff were forthcoming. 'Mam'; my son knew me well. It turned out to be a wonderful experience. I went with them on several occasions and even helped some of them with making some of the craft-based parts of their projects.

My younger daughter also started dancing classes. Most of their costumes were made by a few of the parents. Again, I got involved, again mostly with accessories – such as about 30 crocheted 'ABBA' hats for one show. But all good things come to an end. When she left for university things once again quietened down, but only for a few years.

My elder daughter lived in Wales at that time. One day in about 2009 she phoned me up to say she'd had the best weekend ever! A friend of hers, an event organiser, had helped to start up Wonderwool Wales – a wool festival in mid-Wales, and had recruited her as a steward for three days' work. Apart from one year, when work commitments wouldn't allow, she has been every year since. In 2011 she asked if I'd like to be a volunteer steward. It's a long way to go but we have a touring caravan, so decided to combine it with a holiday. During the time of the show we used the campsite in the grounds of the Welsh National Showground where the festival is held. I was only helping on the two days of the actual show, so on the Saturday morning I had breakfast with my husband and was about to set off. He was settling down to a nice day of rest. Then my daughter arrived. Andy, one of the organisers, had radioed to see if her dad could help out in the car park. Later he helped a stallholder out by standing guard over her stall while she attended a presentation. He was immensely proud to announce he had sold one of her toy sheep! He was hooked! Two years later, our younger daughter happened to be at home and she got roped in as well. This was getting to be a real family affair! After four years my husband was unable to go, but I've been every year since. It's mostly the same stewards each year and it's like a family reunion.

The stewards at Wonderwool wear hi-viz waistcoats, but being the event it is, I got the idea to make my daughter a crochet version (in luminous yellow with sparkly grey stripes), for our second visit. As she is a fan of pink, I also made her a shocking pink one. It was just a bit of fun, but they were well liked so for the next year I made another five – all in different colours. The following year was another five, this time the stripes in contrasting colours. The stewards said they'd never had their pictures taken so much! As so many visitors had been asking about the pattern, I eventually made the decision to write it out, but sold it in aid of a charity (the Lymphoedema Support Group). My daughter and Astrid tested the pattern for me, thus making their own versions. Since then they have both made another but every single one is still unique!

Another great thing about Wonderwool is the 'Sheepwalk'. It's like a catwalk for the exhibitors to display their goods. I think that's what got my daughter so excited about Wonderwool in the first place. By the time I got involved, she was organising it and I was assigned to help her. After another couple of years, she was compering it and I'd taken over most of the organising. When we needed volunteer models, it was great to go into the audience to recruit them. If you ask a group of women who will volunteer, they almost always point to the same person! Children were recruited too and they often turned out to be the most entertaining, plus then they have done something that most of their friends have not.

The reason I am saying all this is because Wonderwool has played such an important part in the gansey project. After all, it was the first public outing of the project!

About 2010 several things happened. Firstly, I retired from paid work, felt sorry for myself for a while, then thought 'get on with it, life changes.' One day, as I was cycling through the local park, there was an event going on where a local councillor had a stall connected with her work at a community centre. We got talking and she invited me to become a member of the newly formed 'Friends of

37

Ridley Park'. I am still involved, helping to make improvements to the park and encouraging its use.

About that time Astrid came to live in Blyth. We had been friends at school and beyond but had gradually drifted apart. She made contact and we realised why we had been friends to start with. Though she still worked full-time, she was looking for something to do and, as the local theatre was in the street where she lived, that was the first port of call. At that time, it was run entirely by volunteers and we offered our services. We started by doing front-of-house duties and taking our turn with the Saturday coffee morning, but soon got more involved. We both helped in the wardrobe department, then props and even a little set making and dressing. Astrid quite soon gravitated towards becoming the props person, but I stayed helping out, mainly with wardrobe and especially accessories.

So, you can see from all this how it was almost inevitable that we got involved with this project. At first, we were only going to knit a gansey each. Even though we'd never knitted one before, we were both experienced knitters. Then came that fateful meeting with Clive Gray, the Chief Executive of Blyth Tall Ship. We were never coerced or cajoled into doing this, in fact we talked ourselves into it! It has been a lot of work (literally thousands of hours – I've been counting!), but I never once wished I'd never started and I don't regret a single minute of it.

Chapter 3

William Smith and his discovery of Antarctica in February 1819

William Smith was born in Seaton Sluice, Northumberland, on 11 October 1790. His father, also named William Smith, was 50 years old when he married Mary Sharp of Earsdon in 1785 (Northumberland County Records Office, PEA5).

William Smith Snr was a joiner from Seaton Sluice when it was a centre of glass making, although other local occupations included blacksmiths, millers and farmers, as well as sailors and ship-masters indicating the scope of the industrial world in which William Smith grew up.

Some time during the 1790s, the family moved around five miles north to Bedlington, probably because there would be more work for Smith Snr there. In 1795 a daughter, Elizabeth, was born (Bedlington Parish Register). Not much is known about William's formative years but, since there were several academies in Blyth, it is thought he might have been sent to school there.

It is likely William Smith was apprenticed to sea at the age of around 14, probably spending much of his time in the East Coast coal trade. This was a developing industry at the time. Blyth becoming a leading port in the trade and was thought to be a good training ground for seamen.

He is also thought to have worked in the Greenland whaling trade. There are no specific records to support this, but there are records showing that 14 ships sailed from Newcastle to the whaling grounds at East Greenland and the Davis Strait. This trade was particularly active from 1804 to 1807.

Smith came out of his time on 27 August 1811 and in the following month it is recorded that he was the master of the *Three Friends of Blakeney*, a ship used in the coal trade. It is not known how he raised the money for the ship but it is quite astonishing that, so early in his career and at the age of only 21, he was master of his own vessel. Although there are no records to support this, it could be that he borrowed from the Ridleys, the local bankers. This possibility is supported by the fact that a small island off the north coast of King George's Island in New South Shetland was named Ridley's Island.

In 1811 to 1812 a new vessel was built for him by his partners William Strangham (rope maker from Blyth in partnership with Mr Davidson) farmers William Sheraton and Edward Storey and George Jeary, a master mariner from King's Lynn. Because three of the chief owners were called William, the ship was named the *Williams*. The ship was built at Messrs Alexander and John Davidson and was registered at Newcastle on 21 January 1812 (Newcastle Custom's House).

Smith then began a series of voyages that took him

farther and farther afield. In July 1812 he sailed from Gravesend for Lisbon returning to the Thames on 27 December. He undertook a further voyage to Lisbon in August 1813 returning some time in November. His next recorded voyage was to and from Bordeaux between October 1814 and January 1815.

At this time, the Spanish colonies were gaining greater freedom and markets were expanding. There was keen competition for this trade. After spending two months unloading and taking on cargo, William Smith sailed for Buenos Aires, returning at the end of December 1815. On 20 March 1816 he left Portsmouth reaching Buenos Aires on 11 June and returned to Gravesend on 29 December.

He sailed from London to Buenos Aires, arriving there on 12 April 1817. On the return passage of his next voyage, the *Williams* was taken by a Spanish letter of marque into Bahia, where Smith landed part of his cargo which was said to have belonged to a Buenos Aires merchant. He was then allowed to proceed.

William Smith had his own version of these events.

Copy of a letter received from Capt. William Smith of the brig *Williams*, from Buenos Ayres round to London:

St Salvador, Bay of All Saints, Coast of Brazil, Dec 1817.

I am sorry to inform you, that in lat. 13.17.S and long. by account 31.00 lat. of Greenwich, on the 13th November, we were captured and plundered of all our valuables by a Spanish letter of marque with a cargo of slaves on board, and sent into port on the 20th November, during which time we have been under quarantine, and myself, together with the crew, having no communication with each other, not being allowed to speak a word of English from the day of our capture, by command of the Spanish Captain. I now have protested, and will do every endeavour, as far as lays in my power, for the benefit of all concerned. The British Counsel has our case laid before the Governor, and I am in hopes of being liberated soon; but in my next I shall be able to give further information; but from the large quantity of specie taken out at sea, and all parts of the ship broke down, I am afraid we shall find a deficiency of money by plunder in the end.

William Smith

On her fourth voyage to South America, the *Williams* left on 29 July 1818 with a varied cargo. After calling at Rio de Janeiro and a passage of 85 days, Smith arrived on 26 October, where he stayed until January 1819. His cargo consisted of cotton silks and hose, wrought iron (valued at £500), hats, tools, apparel, cast iron, cutlery, saddles (valued at £531) musical instruments, music books, wine and confectionery. The *Williams* reached Valparaiso on 11 March 1819 (48 days out from Buenos Aires) with goods from England and India listed.

It was during this voyage that William Smith discovered New South Shetlands (or the South Shetland Islands to use the present name) and left his name in the history of Antarctic discovery.

The discovery would have been authenticated in William Smith's log which would have been handed to the custom house on his return to the Thames in 1821. Sadly, this would have been destroyed after a number of years. A duplicate log may have been kept by William Smith himself, but this has never been found. His original manuscript chart has also been lost, so all that remains is a printed chart which appeared in October 1820 to accompany an article written by John Miers, dated Valparaiso January 1820. Miers was an engineer and scientist of some standing. He arrived in Valparaiso about two months after Smith had sailed on his first Antarctic voyage.

When William Smith arrived in Valparaiso on 11 March 1819, he reported his discovery to Captain William Henry Shirreff of HBM ship, *Andromache*. Shirreff was a senior naval officer on the Pacific coast of South America. He felt that Smith had probably mistaken ice for land and had not taken soundings to support his claims.

On hearing of his account, English merchants (according to Miers) *"all ridiculed the poor man for his fanciful credulity and his deceptive vision."*

Miers went on to say, *"Mr. Smith was not, however, to be thus easily laughed out of his own observation; he… had learned to distinguish land from icebergs; though it must be confessed that the most experienced eye is often deceived by the striking similarity."*

William Smith, in his determination to prove his claim, sailed from Valparaiso on 16 May 1819 intending to visit his newly-discovered land again. However, this was the Antarctic winter and on 15 June, Smith found himself hemmed in by loose ice. He immediately hauled slowly northwards under sail to minimise the damage to his ship. On reaching Montevideo, he found the ice had taken several sheets of copper from the bottom of the ship. This repair must have been expensive and taken some of the profits from his cargo.

He also learnt that news of the discovery had preceded him and that a sealing ship *Espirito Santo* was bound for the islands. On Christmas Day 1819 this boat, commanded by Joseph Herring landed at an unidentifiable place two months after Smith. This ship was the first of many sealing expeditions to visit the New South Shetlands over the coming decades.

Smith spent three months in Montevideo gathering enough cargo for a third voyage to Valparaiso and a visit to his new islands. He was offered considerable sums by American merchants to take a whaling ship south to the new lands but Smith refused, wanting only to land and claim the islands for the British.

In September 1819 Smith sailed once more to Valparaiso and, in cloudy conditions on the 15 October 1819, he sighted land and took soundings, discovering 60 fathoms and fine black sand. The island was most probably Desolation Island.

On 17 October Smith sailed the coastline of King George Island, Nelson Island, Robert Island and Greenwich Island, finally arriving once again at Williams Point. Smith had sailed 150 miles along the coastline and on 18 October he sighted new land to the west south west that was higher than yet discovered - Smith Island at 2,104m.

Finally, back in Valparaiso, Smith's claims were now believed by Captain Shirreff. Smith was prevented from any contact with the shore leading to the rumour that a great discovery had been made. Captain Shirreff chartered the *Williams* for the Royal Navy and installed Edward Bransfield (master of Captain Shirreff's boat the *HBM Andromache*) and three midshipmen, C.E Poynter, Thomas Maine Bone and Patrick J Blake. Dr Adam Young, an assistant surgeon in the sloop *Slaney*, also joined the crew. The expedition's instructions were to:

1 ascertain whether Smith's land was an island or part of a continent and explore it to the eastward, southward or westward according to circumstances
2 explore the harbours and make charts and ascertain the latitudes and longitudes
3 note the sperm whales, otters and seals upon the coast, collect specimens of plants and ensure that Mr Bone made drawings of every animal, bird, fish, insect and reptile
4 note the appearance of the land and collect rock specimens
5 keep meteorological and magnetic records
6 observe the character of the inhabitants
7 take possession on each quarter of the land, separately

An artist's impression of the brig *Williams* at its most southerly position on 23 February 1820 from a painting by Commander G W G Hunt Royal Navy. Courtesy of Richard Campbell.

They were instructed that if the voyage should last more than six months, they should sail directly to England and report to the Admiralty.

Smith and Bransfield sailed south and made land on 19 December 1819 at Cape Shirreff on Livingston Island. They ran along the north shore of New South Shetland to North Foreland and Cape Melville. They sailed on to Deception Island, sighted Trinity Island and behind that the mainland of the Trinity Peninsula, the northern tip of the Antarctic Peninsula and continent. This then was the first recorded sighting of the Antarctic continent in history.

Following this the *Williams* sailed past Elephant Island (made famous by Shackleton's incredible self-rescue) and Clarence Island whereupon shortly afterwards they were halted by sea ice in the Weddell Sea. On their return voyage they ran alongside the north coast of the New South Shetland Islands before finally heading home for Valparaiso.

A daily journal of the voyage written by Midshipman Charles Poynter can be found in a book published by The Hakluyt Society in 2000 called *The Discovery of the South Shetland Islands 1819-20*. The book, edited by R J Campbell, is full of further information including contemporary press reports and copies of letters sent by Captain Shirreff to the Admiralty in December 1819 informing them he is hiring the brig *Williams*, under the direction of Edward Bransfield to survey and chart the new discovery.

Bransfield and Smith arrived back in Valparaiso on 15 April 1820. Bransfield rejoined Captain Shirreff's ship and the *Williams* was discharged from service. The name New South Shetland Islands was officially given to the islands. Smith made a final fifth journey to the area on his return home and arrived in Portsmouth, England on 11 September 1821 via Rio De Janeiro and Lisbon. Upon arrival Smith discovered that his fellow owners were in financial difficulties and that he was in fact bankrupt as a result.

The *Williams* was sold on 13 June 1822 firstly to merchants in Ratcliff then to a London coal merchant. The ship was re-registered in Newcastle upon Tyne in 1833 and was subsequently owned by a Newcastle/Jarrow shipbuilder and then a Darlington coal owner, a Bishopwearmouth and Monkwearmouth shipbuilder, a master mariner and eventually John Furness of West Hartlepool who owned and eventually had the ship broken up on 12 December 1882.

William Smith lived in London after his bankruptcy working as a river pilot on the Thames. In 1827 James Weddell (who ventured to the New South Shetlands after Smith's discovery and subsequently had the Weddell Sea named after him) recruited Smith who became master of a whaling ship out of Leith and spent some years working in the Davis Strait near Greenland. However, after a bad season in 1830, he returned to London and presumably resumed working as a pilot. Increasingly destitute, in 1838 he made an application to an almshouse but was rejected as he was under the required age. Finally, in 1840 Smith was allocated a place in an almshouse and in 1847 he passed away leaving his household goods, effects and stocks to his wife Mary.

Although there are some discrepancies in the reports, there is no doubt that Smith did discover some of the Antarctic islands on the first of his voyages to Antarctica. One of the sources that support this is a memorial written by Smith and presented to the Admiralty on 31 December 1821, shortly after his return to England and written in his own hand, when he was hoping for some reward for his discovery. The Admiralty required confirmation from Captain Shirreff and a letter was sent but no record has been found of any outcome.

William Smith's Memorial to the Admiralty. As the original is held in the National Archives, this is a transcribed copy from Bob Balmer's research. He lists the archive reference as Pro. S 498-1821. ADM 1.

Dated 31 December 1821
To the Right Honorable the Lords Commissioners of his Majesty's Admiralty. The Humble Memorial of William Smith of the Ship WILLIAMS of Blyth -
That your Memorialist on a voyage from Buenos Ayres to Valparaiso in Chile on the 19th of

Feb 1819 – when rounding to the Eastward of the Falkland Islands, the wind prevailing from the West North West to the West South West, and finding no chance of making a passage round Cape Horn, without being in a high Southerly Latitude, keeping a good lookout foe [sic.] ice. When on the 19th aforesaid at 7 a.m. Land or Ice was discovered veering South East by South, distance two or three Leagues – Strong Gales from the South West accompanied with snow or sleet – Wore ship to the Northward, at 10 a.m. more moderate & clear, wore ship to the Southward and made sail for the Land – at 11, rounded a large Ice Berg, at noon, fine and pleasant weather – Latitude by Observation 60 dg 15m South – Longitude by Chronometer 60dg 1 m West – steering in a South South East direction – at 4 p.m. made the land bearing from South South east by East, distance about 10 miles, hove to, having satisfied ourselves of Land hauled to the Westward and made sail on our voyage Valparaiso –

That your Memorialist sailed on the 16th May 1819 – from Valparaiso with a cargo for Monte Video the River Plate, being resolved to visit again the new discovered Land, that on this voyage nothing particular occurred until the 15th of June 1819 – at 6 p.m. being in Latitude 62dg 12m South & Longitude 64dg West, he found the sea very smooth with some small particles of Ice and supposed it might be from a heavy fall of snow; in less than half an hour, we were completely hemmed in by loose Ice, hauled our Wind to the Northward & …regulated sail, so as not to go more than from one to one & half knots through the Ice for fear of damaging the Ship - at Midnight, clear weather & it not being practicable to remain in so high a Latitude at that season of the Year proceeded on the Voyage to Monte Video convinced that land must be near from the forming of Ice.

That on Memorialist's arrival at Monte Video the Report of his Discovery got into Circulation, that the Americans at that Port & Buenos Ayres offered your Memorialist large sums of Money to make known unto them the Discovery he had made, but your Memorialist having the Good of his Country at hart [sic] (if any should be derived from such Discovery) & as he had not taken possession of the land in the name of his Sovereign Lord the King, resisted all the offers from the said Americans, determined again to revisit the new-discovered Land –

That your Memorialist sailed from Monte Video on a Voyage to Valparaiso, with the intention of again looking for land in the high Southern

Latitude = That on the 15th of October at 4 p.m. your Memorialist discovered Land bearing S.E. by E., distance 3 Leagues – Weather hazy, soundings at 65 Fathoms with fine sand & ouze – Wore Ship to the Northward for the Ensuing night – 15th, in the Morn made sail for the Land, at Noon fine & Pleasant Weather observed a large track of land laying in a W. S. W. & N. N. E., direction, very high & covered with Snow, vast quantities of Seals, Whales & Penguins etc., about the ship.

That on the 17th day of October 1819 – your Memorialist landed took formal possession of the new-discovered Land in the Name of His Majesty's George the Third & names the Land New South Britain, & after making every possible discovery made sail for Valparaiso.

That on your Memorialist's arrival at the said Port, he immediately reported a second time his discovery & taking possession of the said Land in His Majesty's name to Captain Sheriff of His Majesty's Ship Andromache, with the different Soundings & all the observations your Memorialist had made, upon which Captain Sheriff was pleased to send the Ship Williams, with sufficient number of Officers & seamen for the purpose of visiting the new-discovered Land named by your Memorialist New South Britain – On the return of the Ship Williams to Valparaiso from the said voyage, she was discharged from the service by Captain Searl of HMS Hyperion & the name of New South Shetland was given to the new-discovered Land.

That your Memorialist sailed from Valparaiso for the Coast of the New South Shetland for the purpose of fishing for Whales & Seals with a crew of 43 men – on your Memorialist's arrival on the said Coast, he found several Bays and Harbours which are pointed out in the Chart submitted to your Lordships with this Memorial, that shortly after to your Memorialist's surprise, there arrived from 15 to 20 British ships, together with about 30 sail of Americans, & during the fishing season it was with great difficulty that your Memorialist maintained Peace between the crews of the two Nations who were on the shore – There are many mineral productions on different parts of the Coast – The Fishing trade has been followed up with great perseverance (sic) & has already employed about 1200 British Seamen. – The rest of the Coast has not yet been ascertained.

Your Memorialist therefore humbly craves your Lordships will be pleased to take into your Consideration the anxious solicitude with which your Memorialist followed up his attempts to *discover & ultimately by his perseverance (sic) discovering the Land named by him New South Britain, his having taken possession of the same in His Majesty's name & through his information to Captain Sheriff of HMS Andromache the same discovery having been fully proved to be true & correct & that your Lordships will be pleased in consequence to grant to your Memorialist such remuneration as in your judgement may seem to meet – Any Your Memorialist as in duty bound will ever Prey (sic)*

No 5 Griffin Street, Shadwell
W. Smith

References

Captain William Smith and the Discovery of the New South Shetlands, A.G.E. JONES, The Geographical Journal, Vol 141, No. 3, Nov 1975

The Discovery of the South Shetlands:

The memorial to William Smith

In November 2019 the Duchess of Northumberland visited the Blyth Tall Ship workshop to unveil a memorial to Captain William Smith.

The memorial was commissioned by Blyth councillor Gordon Webb who used his discretionary fund to pay for it.

Her Grace, The Duchess of Northumberland, has been a patron of the project since its inception. It was a very cold and windy day - but at least the rain stayed off. Her Grace was welcomed to the event by Blyth Mayor, Warren Taylor.

The memorial was designed by Russel and Wendy Taylor from Blyth Bespoke Fabrications. Ashley Brotherton and Andrew Holmes did the artwork.

It's lovely that when you look at the sculpture, the *Williams II* can be seen in the background.

The memorial can be found near to the RNLI centre on Quay Road - next to the river Blyth.

The Duchess was presented with a grey gansey knitted by Meryl Goetsch who lives in Cape Town, South Africa. Meryl was delighted that her gansey had been given to such an auspicious person.

It Started with a Stitch

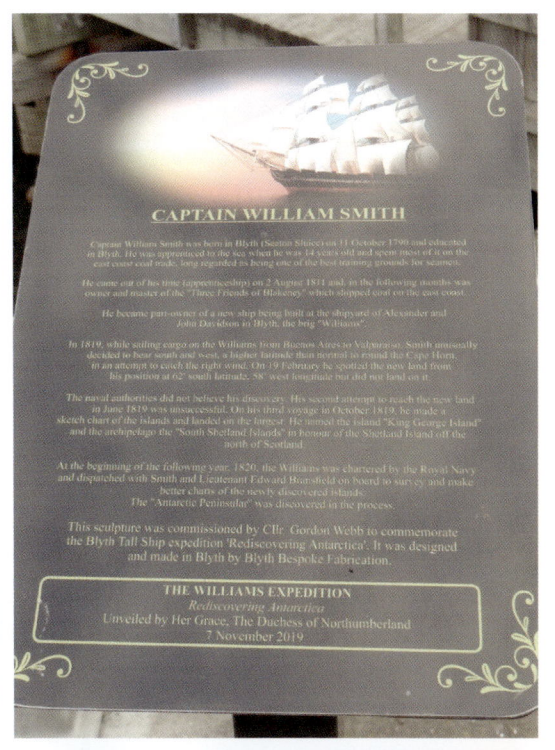

Chapter 4

Blyth Tall Ship Project

Exploring our future, inspired by our past

Williams II in full sail

Back in 2008-9 an innovative experiment was conducted by the Extended Services team from Northumberland County Council working with young people identified as likely to become 'NEET' (not in employment, education or training) to see if working on a project to build a small traditional sailing boat could impact the potential for continuing in education. The young people chosen were thought to have potential but, because they were not engaging with their education, they seemed unlikely to reach that potential.

The project, based in a local school in Blyth, set out to work with this group of young people to build the small boat, introducing the students to new skills that would engage their interest. The project covered the school year 2008 to 2009. A small skiff was made within school premises. This was facilitated by employing a boat builder and a life coach. The project was a great success. Not only did many of these young people benefit from working on a creative project, but many went on to study in the sixth form.

It was an expensive project and reached only a limited number of students. However, encouraged by the success of the project, Northumberland County Council decided to explore the possibility of building a larger ship. Firstly, they held an open meeting to gauge local support for extending and widening the project and given the positive interest, the local authority identified a pot of money to set up a feasibility study into building a wooden Tall Ship from scratch in Blyth to involve a wider group in the project.

They advertised for someone to carry out the feasibility study and Clive Gray, now Blyth Tall Ship CEO, successfully applied for the post that provided one year's funding. Clive worked with a project board including representatives from Newcastle University, Port of Blyth and Blyth Yacht Club. Three months into the study, it was obvious that the cost of the proposed project would be exorbitant. Clive advised the local authority of his findings in a project report including a 10 year forward plan and then suggested that he should use the remaining nine months' funding to look at how he could deliver the 10-year plan. He began to write bids to funders such as the Coalfields Regeneration Trust and the Heritage Lottery Fund (HLF). The first £100 came from Blyth Rotary Club, then the first successful HLF funding was received. Port of Blyth had an old building that had been used for boat repair. The condition of the workshop was poor and there was asbestos in the building. The final cost of the refurbishment of the workshop was around £30,000.

Blyth was still suffering third generation unemployment after the collapse of the coal and shipbuilding industries with many young people struggling to identify with their future and aspire to the opportunities that were developing in the Port of Blyth.

Unfortunately, there is a skills gap between those leaving formal education and these growing businesses.

Clive Gray felt that setting up a workshop to teach heritage skills to young people who had left school with few skills or employment possibilities would be a very positive addition to what was then available in Blyth for these young people.

Cross section of the Zulu

The local authority supported this proposal. The grant from the Heritage Lottery Fund was used to set up the Blyth Tall Ship (BTS) Project. The project began with minimal staff. Clive was appointed as CEO. Paul Allen and Jonny Bell were employed as shipwrights to train NVQ trainees in the workshop.

Young people were taught basic woodworking skills and went on to work on boat building and repair within the workshop. Throughout the project many young people moved on to higher education, apprenticeships and employment. BTS has enjoyed a higher success rate than many other similar projects.

In the beginning, the project was based in one small workshop. With the help of Blyth Port Training Services, the project now has two workshops. The second is a much bigger building with space for more boat building and repair activity.

The light at the end of the tunnel is the emerging offshore and engineering sector that is growing in the area and the expanding Port of Blyth.

The project began delivering one-day curriculum benchmarked schools experiences for 8 to 14 year-olds and foundation learning in engineering skills for 14 to 28 year olds, through the medium of heritage boat building, as an inspiration for change.

They are already supplying newly trained young people as apprentices or into full time jobs in the local engineering and offshore sector with over 20 percent gaining employment and 40 percent taking on further learning.

The Zulu

Blyth Tall Ship Project eventually secured HLF funding to build a wooden masted ship from scratch. The ship (Zulu) is a herring drifter. These ships were built in very large numbers, their speed and excellent seakeeping qualities making them ideal for work in the North Sea. They were built in many different sizes, some up to 60 feet in length. The building of the Zulu has been delayed by the Covid pandemic and by the need to concentrate work on the *Williams II* but is coming on well.

William Smith

During the early years of the project the story of William Smith came to light. William Smith was a Blyth sea captain who discovered the continent of Antarctica in 1819 (see chapter 3). Money was raised to buy a 100-year-old brig (the *Haabet* or *Hope*) which was brought from Denmark for refurbishment. The initial plan was to recreate the voyage of William Smith in 2019, the bicentenary of William Smith's discovery. The ship was renamed the *Williams II* by Sophie, Countess of Wessex, in a ceremony held in April 2016.

So, by combining the heritage experience of reliving Captain William Smith's adventures and working with the growing businesses and the community in Blyth, Blyth Tall Ship Project hopes to see both the community and the unsung hero find new and exciting futures.

Success stories

Nigel Gray

The Blyth Tall Ship Project is peppered with success stories, one of which is the contribution that retired professional rigger, Nigel Gray has made to the project.

Nigel had owned a fishing boat when he was younger and has had an interest in rigging and ropes since he was a child.

After selling his boat, Nigel decided that he wanted to do rigging professionally.

Sixteen years ago, he contacted several companies offering to work for nothing to gain better rigging experience. Work experience of this sort is hard to come by and not available in the North East.

Sharing his time between home and Essex, Nigel was eventually offered work experience in a company in Essex. Living on a barge during the week, he crewed on the barges at weekends.

A big break came when the *Cutty Sark* was undergoing repair and rigging skills were needed. Nigel headed up a team of inexperienced workers to undertake this task. The rigging was taken to a workshop for overhaul and repair and the inexperienced team learnt new skills that in turn increased their feeling of empowerment and confidence.

Through this, Nigel realised that he had a knack for training and that he could break tasks up into manageable chunks so that they were achievable.

Examples of rigging techniques

Nigel Gray

Coming up to his 65th birthday, Nigel decided it was time to retire professionally.

Nigel has a keen interest in social inclusion issues and how these can affect routes into employment. He has also read reports by the Joseph Rowntree Foundation, discovering that white males are one of the most marginalised groups in society. This client group is well represented at BTS.

At the same time, Nigel found out about the Blyth Tall Ship Project and the intention to buy a ship. Nigel wanted to carry on teaching but not only that, he was keen to be involved in a project that was really making a difference in the lives of trainees.

Rigging was incorporated as a module into the NVQ2 so that trainees could become accomplished in different aspects of rigging, which is a transferable skill.

Under Nigel's tutorage the rigging of the *Williams II* has been overhauled. Nigel carried out this work as a volunteer within the BTS Project.

Some of the trainees have gained enough skills to enter into the trade and many others are proficient in rigging basics.

Nigel didn't see his role as simply teaching a skill, but felt it was also addressing other life skills such as personal empowerment and increasing the confidence of the trainees he worked with, making it more likely they would move on to further education, apprenticeships or employment.

There are plans to expand the rigging training at BTS in the future, offering training to outside organisations and groups. Rigging (like gansey knitting) is a heritage skill that is not readily available in the UK. Nigel is very keen to keep rigging skills alive in the UK and feels it would be a great shame if they were all to go abroad.

Williams II entering the River Blyth

Joe Boothby

There are many success stories for trainees who have come through the Blyth Tall Ship workshop. One of these is Joe Boothby who now works on the maintenance team at Port of Blyth. Like many of the trainees who arrive at the BTS workshop, Joe arrived with little or no experience of woodworking or carpentry. During his time at the project, he has learnt many traditional skills and has been involved in quite a few exciting parts of the project.

He began his time at BTS in 2014 when he studied for his NVQ level 1. He had been studying art, so this was quite a change of direction for him.

He gained funding from 2015 to 2016 to be an apprentice shipwright and acknowledges the valuable experiences that became possible thanks to the support he received from the Worshipful Company of Shipwrights.

One of the highlights of his time at BTS was working in Whitby during October and November 2015 while the *Williams II* was out of the water in a floating dry dock. Work was needed on the hull to replace planks below sea level. He had the opportunity to work on re-planking the 100-year-old vessel in Northumbrian oak. The planks were four inches thick and 15 feet long on the stern of the boat, both port and starboard below the water line. He steamed the planks and working quickly, fastened them to the frames in a frantic process that required total concentration, coordination and teamwork. Once the planks were fixed, oakum was used to caulk them in the traditional manner. It was a grueling (but fantastic) month and the sail back to Blyth would demonstrate the success of their work.

Another highlight was when Robson Green visited BTS in March 2016 to feature the work of the BTS Project and the story of William Smith on his Tales of Northumberland programme. He was taken out on the *Williams II* five miles onto the North Sea and even climbed the rigging out at sea. This exposure created interest from local media, boating enthusiasts and local schools, creating a lively and vibrant work environment for a while.

Joe also went to Grimsby with others in the BTS team from September to November 2017 when the *Williams II* needed further work on its hull. It was cold and difficult work and the team didn't get home

from the BTS team has seen timber converted from "in the round" to baulks, and has learnt how to select trees with just the right bend for a particular beam or just the right zigzag to get some grown "knees" (a structural bracket that requires curved grain for strength) out of it.

much during their three-month stay in Grimsby, but the work was needed and the ship was in much better shape for it.

Joe has visited timber yards and with support

Joe proudly made the name boards for the *Williams II* for when she was renamed in 2019.

He has learnt many skills from retired volunteers who visit the BTS workshop, sharing skills, donating tools and talking about maritime heritage. He has gone on to a successful career with Port of Blyth thanks to his experiences at Blyth Tall Ship.

57

Chapter 5

Wonderwool Wales

Wonderwool Wales is the premier wool and natural fibre festival in Wales and is held annually on the last weekend in April at the Royal Welsh Showground, Builth Wells, Powys.

First held in 2006 to promote the market for Welsh wool and add value to products for small wool and fibre producers in Wales, the festival celebrates the green credentials of Welsh wool and its versatility as a material for creative crafts, designer clothes, home furnishings and more. Wonderwool Wales has grown year on year. It covers everything from the start to the end of the creative process – from exhibits of sheep, through raw and hand-dyed fibres, yarn for knitting and crochet, embellishments, equipment, dyes and books, to superb examples of finished textile art, craft, clothing and home furnishings.

Janice's daughter Laura was hired as a steward for the show in its early days when she lived nearby and subsequently persuaded Janice (and even her husband Stuart) to help out. Janice and Stuart have a mobile caravan so towed this down to Builth Wells for the duration of the wool festival.

In 2016, Janice asked one of the main show organisers, Chrissie Menzies if she and Astrid could have a table with information about the Williams Gansey Project. At this time, all they had were ideas and some attempts at the various patterns. This was agreed in return for some stewarding and in April 2016, the very special relationship between Wonderwool Wales and the Williams Gansey Project began.

At that time, the aim of the Blyth Tall Ship Project was to recreate William Smith's discovery of Antarctica in 1819, in 2019, 200 years on. As we now know, various events have prevented that voyage from taking place. Although what Janice and Astrid had to display was very limited, the table attracted a great deal of attention, partly due to the idea of the voyage to Antarctica, but also from knitters who were interested in the Williams Gansey Project.

Deb Gillander from Propagansey offered invaluable advice, especially on sourcing gansey wool. Using this advice, Janice and Astrid bought all their gansey wool from Frangipani based in Cornwall. Jan and Russ Stanland have been unstinting in their support for the Williams Gansey Project for which Janice and Astrid have been extremely grateful.

At Wonderwool Wales in 2016 Janice and Astrid were delighted with the amount of interest and support they received, including offers to knit should funding become available.

By the time April 2017 came around the project had moved on considerably. Funding had been granted the previous November and many of the knitters had been selected from the huge number of offers to knit. Most of the gansey kits had been dispatched. Janice and Astrid actually took some kits with them to Wales to deliver personally to knitters.

One of the highlights of Wonderwool Wales is the sheepwalk. This is a twice daily opportunity for stall-holders to show off their products. Volunteers from amongst the stewards and some hi-jacked members of the audience model garments, scarves, bags etc. Some stall-holders model their own garments.

Janice and Astrid had some finished ganseys to demonstrate and John MacRae, whose wife has the Wool Shack in Malvern, had volunteered to model the gansey and the hat. John, who is a singer, decided to brighten up the event by singing a Williams Gansey sea shanty to the tune of "What shall we do with the drunken sailor", but with the words "What shall we do for the tall ship sailor". He then got the audience to join in a chorus of "Knit, knit, knit a gansey". Amazingly people got involved and sung along.

The other advantage of the sheepwalk was

Janice with John MacRea

Father and son modelling the gansey

that visitors who may not have located a stall they were interested in would, having seen them on the sheepwalk, make a point of visiting the stall later. After each of the sheepwalks, the Williams Gansey Project stall became even more popular. John also visited the gansey stall with his banjo and played a couple of songs.

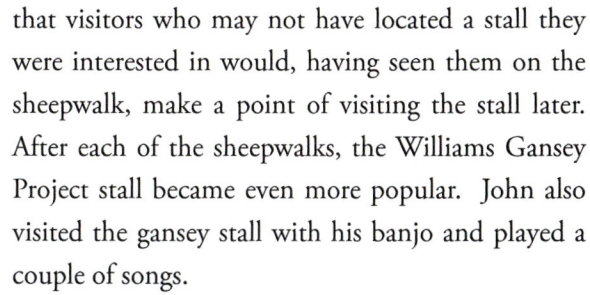

In 2018, the project had moved even further on. Janice and Astrid had more ganseys for the models on the sheepwalk.

By 2019 the pair had raised money using crowdfunding to have grey ganseys and hats knitted for Blyth Tall Ship volunteers who, although they had made a significant contribution to the project, were not going to sail on the *Williams II*. The Round Britain Trials had also begun. Janice and Astrid were able to display some of the group photographs of the various crews on the *Williams II*, along with some photographs and letters from knitters. They also had a map of the world showing where the ganseys and hats had been knitted.

Janice with John MacRea

61

The organisers of Wonderwool Wales have always been extremely supportive and generous to the Williams Gansey Project – giving them space at the exhibition in return for stewarding and a full page in the show guide. Despite a break in 2020 and 2021 because of the Covid pandemic, the Williams Gansey Project hopes to continue its relationship with Wonderwool Wales in the future.

Chapter 6

Woodhorn Exhibition 2018

In Autumn 2017, the Blyth Tall Ship Project was invited to produce an exhibition for Woodhorn Museum, near Ashington, Northumberland, running from January to May 2018. The exhibition was to include the work of the Blyth Tall Ship Project, the story of William Smith and his discovery of Antarctica, the Archive Project (another part of the Blyth Tall Ship Project) and the Williams Gansey Project and was to be displayed in a large exhibition facility.

Woodhorn, a former 19th-century coal mine now hosts a museum and heritage centre for art and local historic relics. It also holds the archive for Northumberland where people can access a wide range of public and personal records.

Janice and Astrid decided they needed to produce a new display for the exhibition to show the progress of the gansey project, however they had very few completed ganseys returned at that time so had to think hard about what the display should contain.

They decided that firstly they needed to plot the development of the gansey project from early planning days to the present time. They also decided it would be good to include something about the history of the gansey and its place in seafaring history. There was also a section on the process of knitting a traditional gansey.

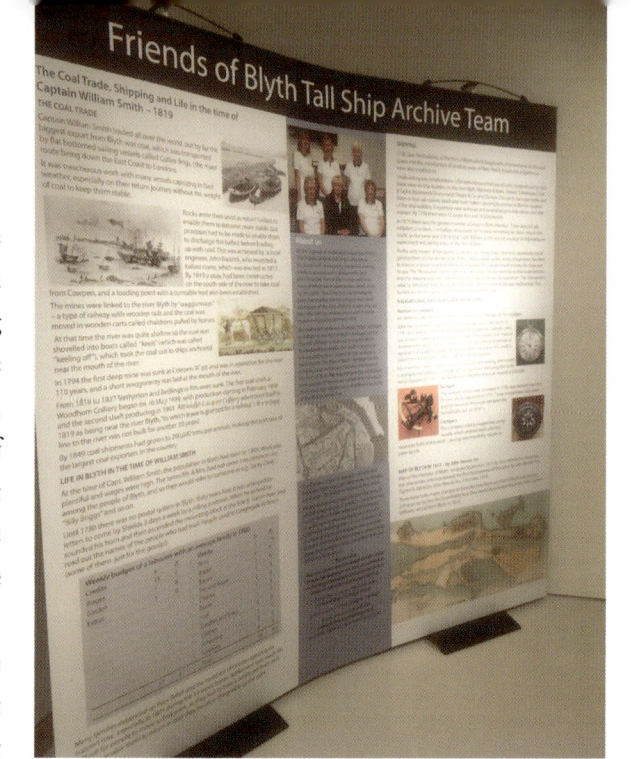

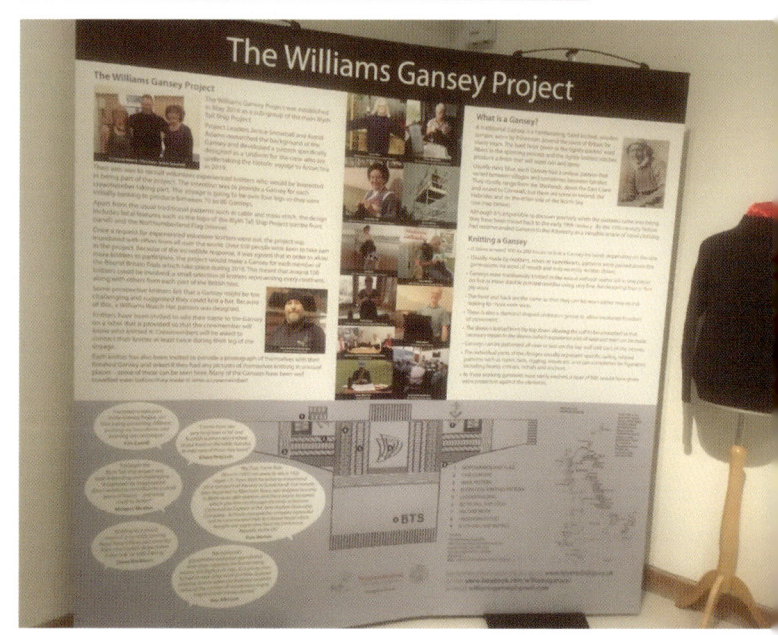

Photographs for the exhibition

The problem of what to do about photographs for the exhibition was solved by Diana Blackburn, one of the Northumberland knitters, who helpfully sent a photograph of herself knitting her gansey on a train from Chicago to Portland, Seattle. This gave the pair the idea to ask knitters to send photographs of themselves knitting their gansey in unusual places.

A number of knitters responded to this request with enough photographs to fill the gap in the exhibition. The photographs ranged from one knitter waiting for her cue to go on stage in a local production of the Mikado to a knitter on fire duties at a festival.

The display also showed the labelled graphic design of the gansey, designed by Janice. Some quotes from the knitters saying why they had volunteered to get involved with the project were also included.

Janice and Astrid were delighted with the finished display which was installed at Woodhorn in January 2018. The gansey that was on display at the exhibition had to be taken there two weeks earlier to be put into the freezer. This was to ensure that no foreign material or bugs were introduced into the museum where other heritage fabric might have been damaged.

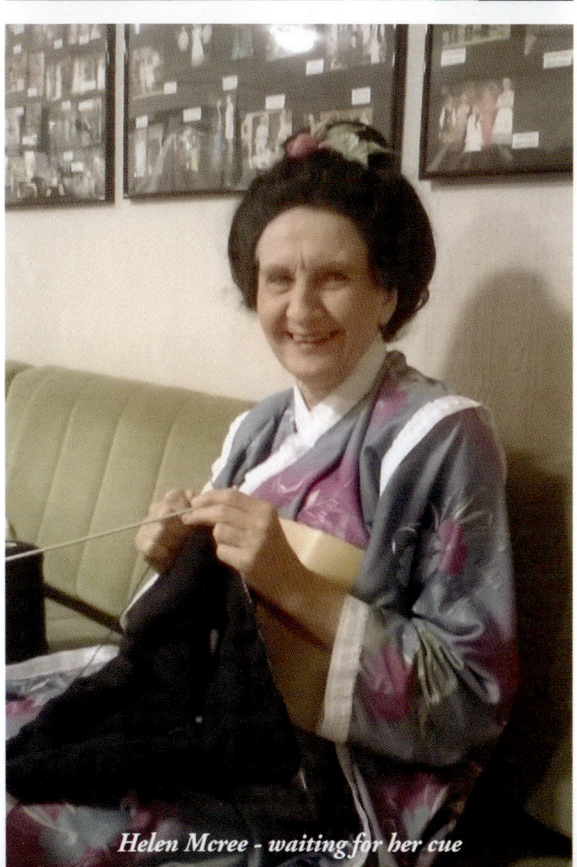

Diana Blackburn on her Canadian train journey

Helen Mcree - waiting for her cue

It Started with a Stitch

Iris Scott - Chilly day on Blyth beach

Maryhelen Toal - Toronto

Anne Shaw - A bit on the big side

Cherry Waters - fire watch duties

Mary Watkins - Who's looking over my shoulder?

It Started with a Stitch

Meryl Herbert - backstage at the Royal Opera House

Philip Batten - Caravan crafting

Above: Tone Carr, a local knitter, took the opportunity to return her completed gansey to Janice and Astrid at the exhibition.

The exhibition room

These photographs show the scale of the whole exhibition, including the front end of *Little Willie*; a small boat made at the Blyth Tall Ship Project by trainees.

The Blyth Tall Ship Archive also had their own display looking at the history of the coal trade in Blyth and the role that William Smith might have had in this trade. There was a section about what the life of William Smith might have been like in Blyth, including the weekly budget of an average labourer's family in 1800.

There was a panel about shipping after the building of the Port of Blyth in 1730 and the establishment of the company of Edmund Hanney of Leith in 1750 and details of navigation equipment available to William Smith during his time as captain.

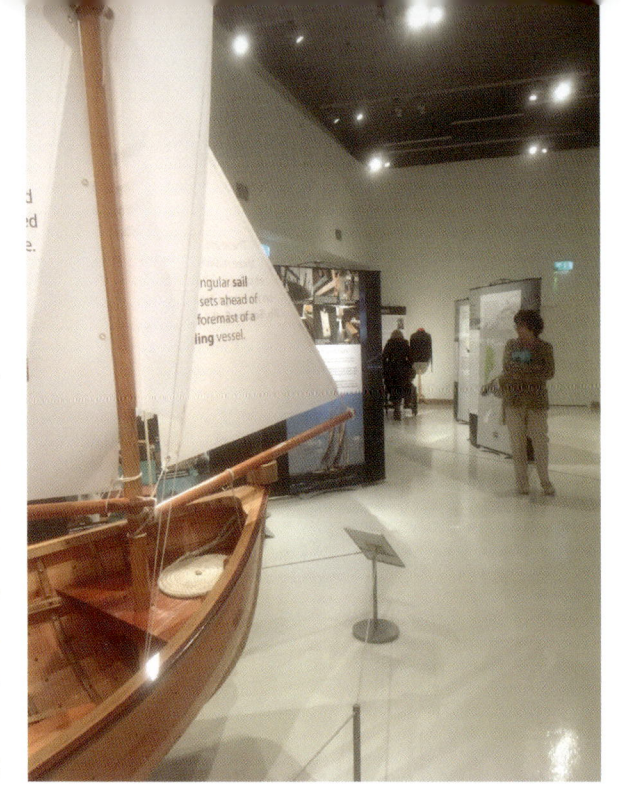

Chapter 7

Knitters' Letters and Pictures

Knitter's letters

Knitters were invited to write to say why they had asked to be involved with the Williams Gansey Project and to send photographs of themselves with their ganseys or hats (if possible). Many of the letters were written when it was still believed that the *Williams II* would make it to Antarctica. This belief is reflected in many of the letters.

This section includes as many of these letters and photographs as could be included.

Megan Allen - Blyth, Northumberland
(Megan has played the Northumbrian pipes on many occasions at Blyth Tall Ship)

> The reason I wanted to knit a hat was because of my involvement with the Blyth Tall Ship Project. Initially I became a piper for the ship, meeting with Clive Gray to play a tune in the Heritage Centre on the quayside and later volunteering - ah, remember those happy days - to clean the ship, literally, from stem to stern. When the gansey project began, I was interested as I considered myself an experienced knitter, but when I learned that a gansey is knitted in one piece, I decided to knit a hat instead as I did want to contribute to the success of the whole venture.

Dear Astrid and Janice,

I'm pretty sure you might have given up on me! I'm so sorry to be such a slow coach.

Here, at last, is my Gansey! What a thrill to knit it for you. I loved the pattern and, especially, the construction method. It was much, much better than either of the two I made previously. I did have a bit of trouble tracking the pattern sheet but an emery board and my row counter helped once I figured out what I needed to do. I added the inch that you requested to the body and sleeves – I hope that I figured them out correctly. I'm sorry about my diary – I started out with the best of intentions but . . . failed.

I'm a transplanted American from Michigan and have been living, with my Scottish husband and two Scottish cats, in Scotland for 27 years. I learned to knit when I was 5 using two ball point pens with the ink removed. My mother couldn't find any short, fat needles at the time. I've enjoyed knitting on and off ever since (50 odd years!). I also crochet, quilt, weave and do a bit of embroidery, although it's not my favourite. I'll send you a photo on line if you really want one! I'm a retired membership secretary for a Scottish charity, an avid, some would say rabid, allotmenteer and enjoy reading and listening to music, learning everything I can about history and would be an avid walker – if I ever find the time!

When my friend sent me your request for knitters I was thrilled with the idea of my work taking part in such an exciting adventure! When you chose me – I was on beyond delighted. I hope that the crew member who receives my efforts finds it warm and up to the job. I apologise in advance for the cat hair that is an inevitable part of all my work – they can just think of it as extra warmth! (Hopefully they won't be allergic! A dry run through a tumble dryer should help.) Please let them know that it has been knitted with love and prayers for safety at sea.

I wish you all the very best fortune in your efforts.

All the Best
Jane

Jane Anderson - Musselburgh, Scotland

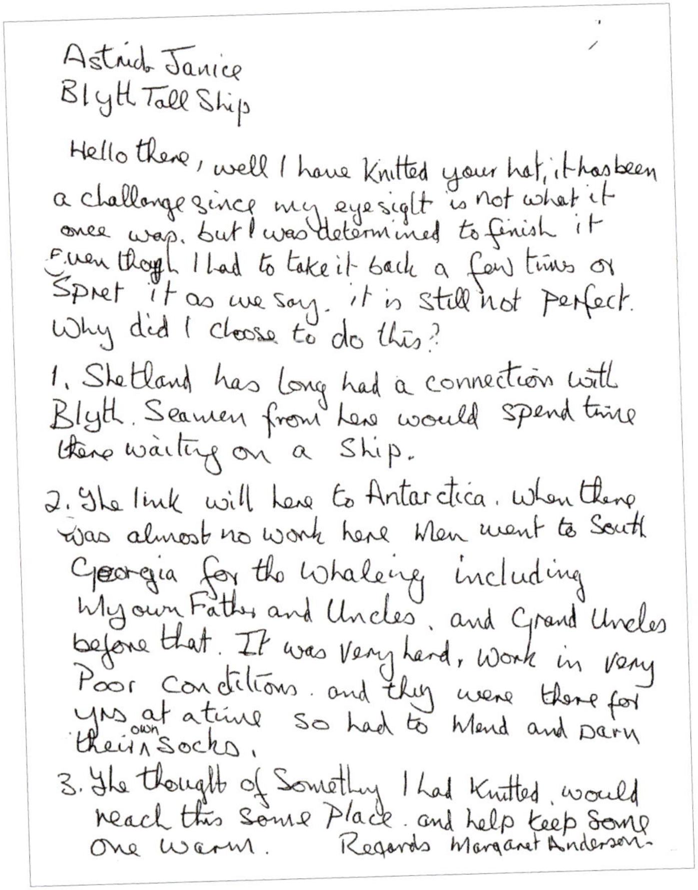

> Astrid, Janice
> Blyth Tall Ship
>
> Hello there, well I have knitted your hat, it has been a challenge since my eyesight is not what it once was, but I was determined to finish it even though I had to take it back a few times or "Spret" it as we say. It is still not perfect. Why did I choose to do this?
>
> 1. Shetland has long had a connection with Blyth. Seamen from here would spend time there waiting on a ship.
>
> 2. The link will here to Antarctica. When there was almost no work here men went to South Georgia for the whaling, including my own Father and Uncles, and Grand Uncles before that. It was very hard work in very poor conditions, and they were there for yrs at a time so had to mend and darn their own socks.
>
> 3. The thought of something I had knitted would reach this some place, and help keep some one warm. Regards Margaret Anderson.

**Margaret Anderson –
Shetland Isles, Scotland**

Doreen Armstrong – Blyth Northumberland

Doreen was a member of the knit and natter group held at Blyth library each Tuesday. She helped us test the gansey pattern by knitting a navy gansey. She went on to knit another navy gansey and a grey one that Astrid was fortunate enough to get to wear when the Williams II returned from the round Britain trials. Very sadly, Doreen died in 2021. She is missed at the library knitting group and her contribution to the Williams Gansey Project was very much valued.

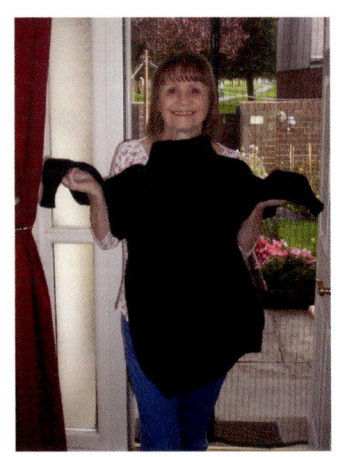

Kay Atkinson – Cramlington, Northumberland

I volunteered to knit a gansey as I have always been interested in knitting more intricate patterns and this was a new challenge. I remember watching my granddad sitting by the fire knitting my socks in feather and fan pattern on four needles though I must have been only been five years old at the time. As a railway signalman he did lots of handcrafting including knitting while waiting for trains but had to retire due to heart problems. My mum and aunt learned from him and always seemed to have knitting 'on the go'. Both later had knitting machines which were used to knit many outfits during my school years. The first garment I knitted was a red four-ply sweater which took so long to knit that I had almost grown out of it by the time it was completed. At an international school camp during my sixth form years I admired the jackets worn by Norwegian contingent. Following the holiday, I discovered just how expensive those jackets were, so I decided to knit one for myself. That proved a problem as I didn't have a pattern; there was no internet in those days. I used graph paper to work out the design and off I went on two needles though I was aware those jackets were knitted on circular needles then cut and bound with ribbon to form an opening. The first jacket I knitted had so many ends to sew in it took ages, but I learnt from my mistakes and the next ones were knitted with the back and fronts knitted in one continuous piece up to the armholes. I grafted front and back at the shoulders with the sleeves being sewn in. Although I have knitted a wide range of items both by hand and on my own knitting machine, including some on circular needles, I had never attempted a gansey.

My husband and I were enjoying a latte at the restaurant alongside the maritime centre when I spotted the leaflet asking for volunteers for this project. I decided to apply as I had some spare time having recently spent available time knitting prayer shawls for church and tiny outfits for the premature baby unit at the RVI (The Royal Victoria Infirmary, Newcastle upon Tyne). This would be a new challenge so, when I was accepted, I was delighted to head off and pick up my kit.

I really appreciated the hundreds of hours Astrid and Janice must have spent creating the pattern for the Blyth Gansey Project now I was to learn how to join the shoulders by knitting a panel by picking up the knitting stitches together as I proceeded. Making a gusset under each arm was also a new innovation for me. The saying "you're never too old to learn" was certainly true for me and I was so proud to complete that first gansey in about seven weeks. I offered to help further with the project and completed a matching hat as well as mentoring a couple of lovely ladies who have needed very little help as the knitting instructions were very clear, more a case of checking details as they went. Recently I have completed a second gansey and found the gussets and shoulder joins easier second time around. If I decide on knitting another Norwegian jacket perhaps I'll incorporate them in my design.

While on holiday in Shetland last year and because I'd become involved, I enjoyed researching the knitting of sweaters and ganseys on the islands then going to a "Makkin' and Yakkin'" session held in the knitting museum in Lerwick where there were some beautiful examples on display.

My husband's grandfather and great grandfather were ships' captains; the former being master of a Blyth-built ship. As a young man he had to seek other work at a New Zealand whaling station when the Krakatoa eruption reduced the number of vessels returning to England until the sky cleared. The family are interested in the history of ships. When the Tall Ships were in Blyth in 2016 our grandson undertook a school project about them. We took him to the Woodhorn Archives where he was able to examine the logbook of the Daisy, a 467-ton barque that carried Durham coal around the Cape to Penang returning with valuable hardwood and other cargo. Unfortunately for the family, the logbook finished in October 1857 the day before his three times great grandfather boarded her for his voyage. It was however fascinating to read all the details of weather conditions and the birds and sea creatures which were encountered. He was able to use the logged coordinates to plot the Daisy's voyage. The Rime of the Ancient Mariner would have made far more sense to me at school had I seen the actual entries in ships' logs!

About 40 years ago my brother-in-law wrote a book about the building of Viking ships as he too had inherited his grandfather's love of ships. It was lovely to see copies on sale at several museums when we visited. It seems now quite dated yet nevertheless accurate. It's fascinating to realise the methods used were very similar to those now being taught in the Blyth workshops. Those same traditional skills were being used when we visited Roskilde in Denmark where a ship was being built to reenact a Viking voyage towards America. Again, they were on view at Lelystad in the Netherlands where a replica of the 1628 ship Batavia was being constructed when we visited. Perhaps the workshop in Blyth could somehow be linked to add another experience to the visitors' centre. Yes, I realise all the health and safety etc. implications but yours is such a worthwhile project which too few people know about.

Another family connection to ship building was when our son worked for Akzonobel who manufacture the paint; yes, I even learned some of the jargon! I have so many links to maritime history through my marriage, so I'm now so delighted to be involved myself.

If one of my ganseys happens to be chosen to be aboard the Williams II, we will look forward to hearing from the crew en route. Life aboard may not have changed much in 200 years but fortunately the technology for navigation and communication certainly has!

Kay Atkinson
May 2017

Linda Barber – St Austell, Cornwall

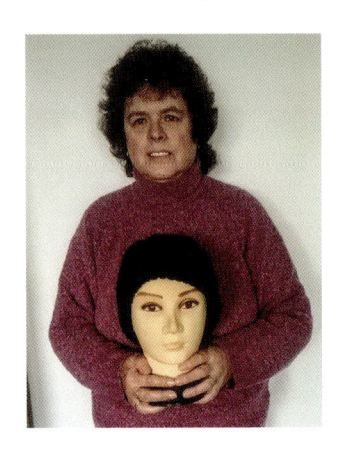

A little bit of info about me. My name is Linda Barber and I am 65 years old. I have been knitting for many years. Several years ago, I used to knit Aran and Guernsey sweaters and sell them at local craft markets. Now I knit mostly hats and gloves (plus a few other bits and pieces). I attend an art and craft group in Gorran where people can pop in to see the various artists and crafters at work and also buy things. I am also learning to paint in oils and I have been telling my art teacher all about the Blyth project because he makes detailed miniatures of ships and is going to look it up online, hoping he will try the Williams. I live in a small fishing village, Gorran Haven in Cornwall, near Mevagissey. From my lounge window I can see the sea and watch the ships and boats go past. Thank you for letting me be a part of this project. I am hoping to go up to watch the crew set off.

Sharon Barclay – Neath, West Glamorgan

Hi, my name is Sharon Barclay and I live in Neath, South Wales. I came across your stand in Wonderwool Wales where I applied to volunteer as a knitter for the ganseys. A gansey is something I have always wanted to knit (I have no sailing background). I have really enjoyed this project as I am always up for a challenge. The only thing I disliked was the needle size. My grandson who is nine loves it and wants me to make him one!! I said maybe in a few months time I will design one for him. I felt very privileged to be chosen and look forward to any updates.

Mary Barnes – France

I heard about the project from a friend in our knitting group who is knitting a Gansey and was inspired by your project. My husband and I are now retired and live in France. When we were first married we decided to build a boat and live on it. We built a 26ft sailing boat and lived on it for a while but eventually work commitments and family meant that we moved into a house. We lived in London at that time and were lucky enough to see the tall ships when they sailed up the Thames. We loved sailing and but we were not quite adventurous as the William Gansey project. We sailed many times from London to the Solent and further afield to Cornwall, Channel Islands and France. Our gansey sweaters served us well for many years of sailing. The Gansey project is a lovely way to include the more sedentary and less courageous of us in this amazing venture. We wish you well with this project and safe sailing and hope we are able to see the mobile project in due course.
Mary Barnes

**Barbara Bignell
Thurmaston, Leicester**

Firstly, may I say what a privilege it has been to be involved in this special project. It's been wonderful to watch the work growing (albeit very slowly at times!) and I'm thrilled to know that it might help keep the cold out at some point in the Williams II's journey. Secondly, may I apologise for not reading my instructions regarding the label correctly and (a) not using a different pen and (b) adding the size when you specifically told me not to!

As a former Wren, this has been a very special year as it is the centenary of the formation of the Women's Royal Naval Service, so making the gansey has given me a real highlight to remember in 2017. We've been celebrating in lots of ways and getting together with friends we knew while serving as well as lots of ladies we'd never met but who all have a common bond, no matter when we joined up or where we were based.

My husband was in the Royal Navy too, a submariner for much of his service, who loved the traditional guernsey I gave him for Christmas one year almost as much as his submarine sweater. He died over 15 years ago but would have been thrilled to know about this project and would, I'm sure, have been happy to model the gansey!

I'm really looking forward to hearing from whoever wears my gansey, as well as news of anywhere it might be possible to see more about the amazing BTS project.

Diana Blackburn
Cramlington, Northumberland

The Gansey Project

My inspiration for knitting the gansey in this project is two fold:-

In memory of my Dad, Charles White, who served on convoy duty in the Royal Navy in World War 2. His ship was H.M.S Kildride. He claimed he was never warm after his Atlantic journeys, maybe a gansey would have helped!

Secondly, as a proud Mum of a currently serving Royal Navy Officer, just returned from nine months detachment in the Gulf on HMS Daring.

It has been a wonderful project to be involved with. I hope this gansey keeps a sailor warm and brings him safely home again.

Diana Blackburn.
Cramlington
Northumberland.

Jill Blunn – Cornwall

My name is Jill and I am 74 years old. I love knitting and this is the fifth gansey that I have made. I love the designs that make it individual. Where Blyth is in the North East of the country, I live right down in the South West only 20 miles from Lands End. Down here ganseys are called guernseys or, if they are plain knitted, they are called knit frocks. Each fishing village would have had its own design, or family design or special to a ship as yours is. I have lots of books of patterns from all over the country. The last one I made was a design that came from Musselburgh. On your gansey there are the sails in the centre with ropes and ladders and flags on the sides with an anchor on the sleeve with the flag of Blyth.

Joyce Brison – Alnwick, Northumberland

I love a challenge and knitting a gansey for Blyth Tall Ship looked like a challenge. I was so pleased to be chosen as a knitter and couldn't wait to get started. I am delighted with the end result and am very proud to have been part of the project - and yes, at times it was a challenge.

I volunteered to knit a Gansey for the project as it seemed like such a lovely thing to do. It's years since I knitted any Guernsey's (very similar) and I relished the challenge of polishing off those skills after years of concentrating on Shetland Lace knitting. It is amazing to think that something I have made will be worn by a sailor on the adventure of a lifetime. I hope it keeps them warm on a chilly night watch. It has been a pleasure to knit the Gansey and be involved, in a small way, in a historic project. My family and friends had now better stand by for gifts of Ganseys as my interest in knitting this style of jumper has been very much re-invigorated :)

Gillian M Britton.

Gillian Britton
Gosport, Hampshire

Jennifer Bryson - Earlston, Berwickshire

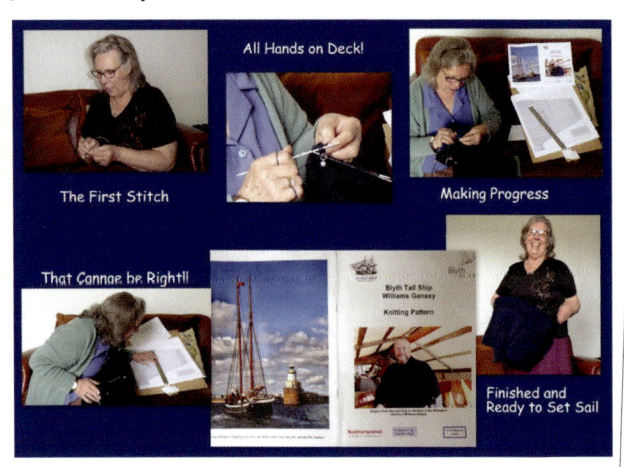

The First Stitch
All Hands on Deck!
Making Progress
That Cannae be Right!!
Finished and Ready to Set Sail

Blyth Tall Ship Williams Gansey Project

Having grown up in a country family post war, I knew nothing other than a family who foraged, for fruit or fuel and spent the rest of their time sewing or knitting.

I was encouraged by a very talented Mum to follow in her footsteps. I do not remember a time of sitting idle, and came to love crafts. As I have grown older I feel my love of knitting has become all important.

When I retired 10 years ago I decided to challenge myself annually. My first challenge was to enter the craft section of The Royal Highland Show and there was a class for a Gansey. I had books on this subject but no pattern and so created one and sourced the yarn to produce a grand jumper for my husband.

When I heard of this project the excitement of the whole idea inspired me to apply. I am so honoured to have been given the opportunity to knit for the sailor and I do hope they enjoy their experience half as much as I enjoyed knitting for them.

Good luck and best wishes

Jennifer Bryson

One of the ladies at my knitting group was knitting a sweater for the gansey project and told me about the Williams Tall Ship. When she said you were looking for knitters for hats, three of us decided we could manage that! It's great to think a hat knitted by me could be in Antarctica! Best wishes for a successful voyage.

Helen Cessford Jedburgh

Linda Coates
Thornhill, Dumfries

Dear Janice & Astrid,

Thankyou for giving me the opportunity to knit a watch hat.

I found the hat quick and easy to knit. I enjoy knitting and like to help when and if I can. I suffer with anxiety and agoraphobia so don't go far from home and it is nice to think that in a tiny way I can be a small part of an Adventure.

Also my husband's late mother is from the Blyth Area. She served as a degauser in the Wrens during the war (WW2) She did her service over in the Northumberland Area.

Hope you think the hat is okay.

Thanks Again

Linda. x

P.S. Would be happy to knit another one. x

Sheena Cooper
Blyth, Northumberland

My first reaction when Janice and Astrid told me about the gansey project was 'Wow, what a brilliant idea!' Many months and stitches later, the gansey is complete, and I'm being asked for my thoughts on the experience.

As I was knitting, I found myself wondering about who would end up wearing the finished garment. I have no idea who you are, but there are many good wishes included in the patterns, and prayers for a safe journey.

I also found my thoughts going back to my mam, who taught me to knit many years ago, and was always the person I would turn to when I got stuck with something.

I suppose, for me, I've found the experience a bit like knitting a bridge between the past and the future, and yes, I still think it was a brilliant idea!

Barbara Coulson – Heddon-on-the-Wall, Northumberland

When I saw the request for gansey knitters in the Northumberland County Council staff e-newsletter I thought 'I can do that.' So, I emailed you and was so pleased that I was chosen to knit one. It was quite exciting when the kit arrived and the knitting started really well. Apart from a slight hiccup over the winter months when I had an excruciating bout of 'knitters' elbow'. My fellow 'Knit & natterers' at Heddon Library were always asking how it's coming along and I think there was a time when they thought they'd never see it finished. It seemed like everyone I met during that time said to me 'how's the gansey coming along?' However, I battled on regardless and am so pleased to have finished something I am very proud of, especially as my gansey may be going half way round the world. It needed a good deal of concentration sometimes and it was a real challenge but seeing the finished article made it all worthwhile. Not sure I'll be rushing to knit another one any time soon, but maybe one day! It's a really interesting project for the North East and I'm looking forward to following the voyage online when it happens. I wish everyone involved all the very best.

I suppose, like many people from coastal regions, I've always had a great affinity with the sea. My great-grandmother, who was one of the Cullercoats fisher-women back in the day, would no doubt have been able to knit one of these in half the time I did, but I'd like to think she would have been proud of me too! Here's a little of her story.

She (Emma Todd) was born in Cornwall in 1869 and travelled to the North East when she was very young. She started married life in Cullercoats in 1893 and was regarded as an outsider for many years. Her husband was a bricklayer but he had an accident and could no longer work, so she became a fishwife so that she could feed her family - she had seven children (no social security in those days!). She operated mainly in the Seghill area, having to go down to the fish quay to buy the fish to sell it. There was no transport in those days, so they had to walk there and then bid for the fish. Times were hard then and she had to share washing facilities with neighbours so she could wash and gut the fish, some of which was kept back to make fish-cakes. Then off she would go to sell her fish, with her creel on her back, even doing this while she was pregnant. Whatever fish she had left she gave away to poorer families. Her fish gown ('goon') was a heavy navy blue material, made by a Cullercoats woman; the skirt had tucks in it, the more tucks it had the richer you were, as it took more material. Then a large apron with a pocket for the money, a shawl, a bonnet and thick stockings. A sharp knife had to be carried to cut the fish. They did have a 'walking-out' gown, usually cotton floral, which they would wear over the top of their navy gown. She died aged 96, having been out shopping that very day.

My name is Kim Cowell, I am 60 years old, and have been knitting for as long as I can remember. Over the years I have knitted many items of clothing but never attempted a Gansey in the round! for me, this project has tested every fibre of my knitting ability, from circular needles to chart work. I found it very challenging, but to see the finished product, I feel very proud to have managed it. I feel honoured to have been chosen to take part. Thank you, and I wish everyone who has / will partake of the voyages success and happiness.

Regards Kim Cowell

Kim Cowell - Kings Lynn, Norfolk

Felicity Denby – London

I am very glad to be part of this intriguing project and to be involved in creating something as useful and as practical as a hat. To think of the distance this object will travel and the means by which it will reach its destination takes my breath away.

Catriona Dickson
Perth, Scotland

Hiya – Just a note to thank you for giving me the honour of knitting this gansey jumper. It challenged me but the pattern was easy to follow. The reason I applied to knit this for you is this – my youngest son had an accident when he was 12. He severely injured his knee, bones, tendons and muscles. As he grew this affected many of his wishes and dreams, but he never gave up. Whilst doing his Duke of Edinburgh (D of E) award he was unable to undertake the adventure part – he tried and failed often. This is where his sailing came to help. Being a young member of the Perth club, he asked and was allowed to sail as the adventure part of the D of E. He went on to sail around the Arctic Circle as part of his gold D of E. He volunteered with the Tall Ships. He trained at HMS Raleigh and has gone on as a sea cadet to escort our Royal family. Now as a mature young man (28 years old) he has moved down to Eastleigh. He continues to sail with the Ocean Youth Trust on the Challengers and is still getting more RYA qualifications. The opportunities he has been given and has volunteered for have given him not only the above but also confidence and new dreams. Me volunteering to knit is my very small way of paying society back. Not only has this challenged me, it has also educated (a bit). I had not heard of Williams Gansey nor the Blyth Trust. I now have.

I personally have arthritis in all my joints so feel I have taken longer than I should have to finish. Thank you very much and I hope the expedition is a success.

Persephone Ditzel – Brunswick, Maine USA

Mother and daughter Persephone and Cassandra Ditzel knitted a hat and a gansey respectively. Cassandra lives in London, while mum Persephone lives in Maine. Persephone writes -

I'm so proud of myself! I finished, and well ahead of schedule to boot!! Thank you for allowing me to be part of this project. I am thrilled to know I am part of it.

You were kind enough to send a kit in care of my daughter (Cassandra Ditzel) who lives in London. She is working on one of the sweaters. She has less free time than I do and a bigger project so is taking a little longer. I am adding some photographs taken along the way of this knitting and I will explain as you come to them. The wool and hat in its various stages has already travelled quite the distance.

The wool was sent to Maine but travelled back to England where the first photo was taken at Stonehenge.

On board the Island Sky on my eagerly awaited expedition cruise to - The Falklands, South Georgia, Elephant Island and the Antarctic Peninsula! Gauge swatch begun - then completed on Saunders Island, The Falklands.

On board the ship was ornithologist Jim Wilson, from Ireland. It was a surprise to him to learn of my knitting project so I told him all about the Blyth Tall Ship Williams Gansey Project. He was on a committee working to erect a memorial to Edward Bransfield in Ballinacurra, Ireland. What a coincidence that he should be one of the guides on our trip!

Thanks again for allowing me the privilege of being part of this project. My father is from Berwick upon Tweed and I have spent quite a bit of time there over the years. My parents have also been several times to Antarctica so for me it's a natural pull to be part of this project.

I've really enjoyed being part of this project; after the initial disappointment of not being selected to knit a gansey, I was pleased to be able to contribute to the project by knitting a hat for a sailor. The ganseys are wonderful and will be treasured by those who wear them I'm sure. My friend knitted one and if I'm honest, I was secretly relieved to be 'just' knitting a hat when I saw how much work was involved in creating the gansey!

Anyway, I hope this hat keeps you warm whoever you are. I have included a little bit of good luck inside your hat - have you found it?

I'd love to hear from you on you travels when 'signals' allow.

I'll leave you with a little Irish blessing.

May the road rise up to meet you.
May the wind be always be at your back.
May the sun shine warm upon your face.
And the rain fall soft upon your fields.
And until we meet again
May God hold you in the hollow of his hand.
Wishing you safe sailing and a wonderful adventure.

Angela Docherty
Hexham, Northumberland

Janet Dunn
Bedlington, Northumberland

I decided to volunteer to knit a gansey because as a knitaholic I like a challenge and this certainly was. Also my husband's family have a history with Blyth ships. Although not sailors, both my late father-in-law and his grandfather worked in both Blyth and the Tyne (Wallsend) shipyards. My father-in-law was foreman blacksmith at Blyth until it closed in the late 1960s. He still maintained an interest in ships, especially sailing ships and painted many that docked at both the Tyne and Blyth docks.

His grandfather came from a humble background but rose to be manager of Bolckows ship breakers at Blyth, then manager at Clellands shipbuilders on the Tyne.

It was certainly a bit of a challenge for me. I didn't expect it to take me so long to complete my gansey. I found it heavy at times to handle and because of the colour difficult on the eyes. I felt a bit sad at handing it over but might make another for myself but in a lighter colour.

I hope you enjoy your challenge as I have mine.
Good Luck

Williams Gansey Project

Hi There

I have had such fun making your hat, there are a couple of small mistakes but it is exclusively yours !!

The reason that I wanted to make your hat is hat I have always been interested in such expeditions and to be involved is such a privilege.

Good luck and I hope you enjoy both the experience and the hat.

Val
xx   xx

Val Donaldson
Northumberland

I received my kit while visiting in Cape Town, South Africa. I have travelled a lot. Unfortunately, I never took the list with me to write the hours taken to knit the gansey. I travelled in South Africa and my knitting went all over from Eastern Cape Province to Gantang in Transvaal and Bloemfontein in Free State. It was the most wonderful project I could ever be part of and I enjoyed every moment. Sorry about the label - forgot to sew it in. Daughter from Cape Town posted it for me. Thanks a million. Joeye Esterhuizen, South Africa.

Joeye Esterhuizen
Noorsekloof, South Africa

Lynne Evans
Carnforth, Lancashire

Dear Janice & Astrid,

Thank you for letting me knit a hat for you. I usually knit Fair Isle hats and thought this would make a nice change. It was more tricky than I thought it would be, and if I wasn't concentrating I found it easy to make mistakes. There was quite a bit of unpicking went on! Colourwork is a bit more obvious.

Anyway, it's finished now! And the labels sewn in. Sorry it's taken so long.

With best wishes

Lynne

I could probably knit another one, but I've got some orders to catch up on first. Let me know if you'd like another.

I've been a keen knitter for years. In January 2017 I wound-up my one-woman gardening business and was open to taking on new projects. When I read a tweet about the Blyth Tall Ship project and the Williams Gansey I immediately felt drawn to give it a try. After all, I'd never knitted a gansey before, I live in the South West of England and have no connection with Northumberland, sailing or the sea. What could be more natural? But I felt that the Williams Expedition was an extraordinary endeavour and something in which I'd like to play a small part.

I picked up the wool for my gansey at Wonderwool in mid-Wales in April 2017. I've really enjoyed discovering the stitch patterns and the construction of the gansey. But it hasn't all been plain sailing. I had to unravel and reknit one section because I'd misinterpreted the pattern. And when I finished the gansey in October 2017 I noticed that one of the cables on the shoulder was twisted the wrong way. I realised that, if only for my own satisfaction, I'd have to fix it; but the gansey got stowed away for months until I felt brave enough to open up the shoulder panel and do the repair. Finally, in February 2018 I was ready to send my gansey off on the next stage of its travels.

I am very glad to have been involved. It's taken me over a year in elapsed time from conception to completion and around 180 hours of working on the gansey itself. In that time the gansey-in-progress has already accompanied me on a weekend trip to Coniston in the Lake District and spent a week aboard a canal boat in Oxfordshire. I wish it well on all its future travels. Even if it doesn't actually reach Antarctica I hope it will have its own adventures along the way. Good luck to the Williams II and all who sail in her.

MAZZY
(mazknitter on Ravelry)

Kate Ferguson – Morpeth, Northumberland

Mazzy Faultey - Bristol

I was delighted to be chosen to knit one of the ganseys for the Blyth Tall Ship Williams voyage to the Antarctic in 2019. When the kit arrived, I was rather daunted to discover I was to knit the largest size of 48 to 50 inches! It did take me some time to finish and I found it fairly challenging, not least because I broke my wrist during the time I was making the gansey, which meant I couldn't knit for two months and was then very slow until my wrist got some strength back! However, it has been a very enjoyable process and I hope that everybody who has taken part in the voyage will find it as rewarding as making the gansey has been.

January 2018

I am an embroiderer based in the Scottish Borders and also enjoy knitting – often hats for the Salvation Army or baby items for our local maternity unit. I am looking forward to following the progress of the expedition and hearing from my hat!

Susan M Finlayson

Susan Finlayson – Hawick, Scottish Borders

Clare Griffel – Bristol

Knitting the Williams Gansey

Back at the start of 2017, my local (Bath) branch of the Guild circulated a message which immediately caught my attention. A tall ship will be sailing from Blythe in Northumberland to the Antarctic in 2019, recreating a voyage by the vessel 'Williams' which made the first sighting of Antarctica, and which took place two centuries earlier. Two enterprising local knitters, Astrid Adams and Janice Snowball, had designed a special gansey to be worn by all the crew members, and had called it 'The Williams Gansey' after the ship. Now they were looking for volunteers to knit the ganseys, with the incentive that the knitter's e-mail address would be on a label sewn into the finished item, and the lucky sailor who wore it would send back information about the voyage – a very clever idea.

As I've had a lifelong interest in the Antarctic, and as I had just finished knitting a gansey for my husband, this was an opportunity I couldn't resist. I had no idea how many people would volunteer, but I sent off my 'expression of interest', and in due course was delighted to hear that I'd been selected to take part. The project was extremely well organised – I received a parcel with enough lovely navy Frangipani gansey yarn, the pattern and charts, labels, a bag for returning the finished gansey, and full instructions. The pattern made use of traditional motifs such as ladders, ropes, moss stitch 'cobbles' and so on, but also incorporated several special symbols including the Northumberland flag, sails and the initials BTS (for Blythe Tall Ship).

I was under no illusions about the amount of work which would be involved – 'my' sailor was a big chap, so there were a lot of stitches in one round! – plus following charts for motifs which have no regular pattern, like the lettering, slows things down. There was a certain amount of swearing at times – for example, when I started knitting back-and-forth for the yoke and forgot, for a couple of rows, that I therefore needed to read the chart in the opposite direction on every other row! And as always, my least favourite part of the process was picking up the stitches for the sleeves – one of these days I must go to a workshop on 'pick up and knit'. Of course, with such a huge project – the charts for each size were different – there was the odd mistake in the instructions, but nothing which an experienced knitter wouldn't be able to correct – Astrid and Janice had done a terrific job.

I cast on just before Easter, and cast off with a great sense of achievement at the end of August (though of course like every keen knitter I'd completed a few other projects at the same time). I waved my gansey on its way to Blythe, and now I'm looking forward to hearing from the lucky wearer – I hope it keeps him/her warm in those chilly waters!

Clare Griffel

Slipknot magazine article

Halee Grimes – Baltimore MD, USA

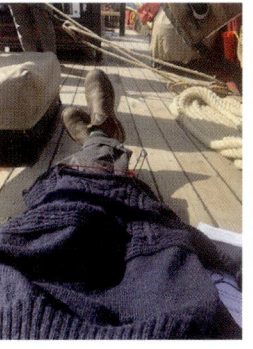

Some pics of me working on the gansey during our transit from Seattle to Olympia. I'm currently chief mate of the Lady Washington. We're making our way through the Puget Sound, Washington. The gansey will have accrued plenty of sea miles by the time it makes its voyage with you!

---

18 August, 2018.

Dear Janice and Astrid,

When I first saw that this project was happening, I was very pleased to get involved. I think the retracing of the route of Captain William Smyth is an inspired idea, and the opportunity to contribute to the project through knitting really appealed to me. I am an avid knitter for more than 30 years, and I love to do something new! By the way, I was really impressed by the package of wool pattern and other material that you sent me.... You thought of everything... and I loved the Frangipani yarn.... A pleasure to work with such wonderful quality authentic gansey wool.

I had knit a gansey before, so I had a fair idea what to it would be like, but it took me much longer to knit this gansey than I had expected! I have a very loose tension, and as a result I had trouble getting the gauge to work out.... my first few attempts were absolutely enormous – more like a dress than a jumper. I eventually went down to a 2.5mm needle, and followed the pattern for a size 40-42" gansey rather than the 48-50 that I had been allocated. The resulting gansey is 52 inches wide – which is still quite large, and should work for a man with a 48" chest. I then had a bit of a tussle with the sleeves.... Solved by reducing the needle size yet again (down to 2.25mm) for the stocking stitch section. I am now quite pleased with the results, I asked my work colleague Philip to model it for me, so that I can see what it looks like on a real living person! I enclose a picture of him wearing it, and of me knitting it!

Overall, this was a challenging but very enjoyable knitting experience. I will be following the progress of the project on your Facebook page with great interest. While I often wished I knew somebody else in Ireland who was knitting a gansey so that we could meet up and compare notes, I really appreciated the opportunity to participate in the project and join a world-wide community of knitters – I love to think that my gansey will be worn by a member of the team who are crewing the Williams II - it makes me feel like I am part of the adventure... thank you so much Janice and Astrid!!!

Yours sincerely,

Eithne Guilfoyle

Eithne Guilfoyle – Dublin, Ireland

**Devon Guthrie – California, USA
(An Opera Singer who spent three years at English Opera)**

To Astrid and Janice, the whole team on the Williams Expedition and the recipient of the gansey. Thank you for the opportunity to participate in such a great project! This gansey was a pleasure to knit and I am excited to follow the progress of the expedition over the coming year. How exciting! Already this gansey has been on quite a journey. I happen to travel frequently for my job as an opera singer (my career actually started at English National Opera!) Along with me on my operatic travelling this gansey has been knitted on countless airplanes, in rehearsals, backstage, in New York, Santa Fe, Chicago, St Louis, New Jersey, Detroit, Kentucky and here in my native California. And now it journeys onward!. So, thank you again. It is a privilege to have my stitches be a very small part of a very big legacy. Best wishes.

Claire Harrison – Gateshead, Tyne and Wear

In 2016 I had found out that the job I had done for almost 17 years was about to be scrapped. I loved the job, even though or perhaps because, it was often very challenging. I was school librarian in a secondary school, which served a difficult area and had high numbers of students from deprived backgrounds with special needs or for whom English was their second language. We'd been made to become an academy and it had been a very pressured time. That same year I lost my father and my favourite aunt. I knew I wouldn't be able to find another job easily at 61 and was feeling very low, I felt I had no real place outside of home. I have a son in his thirties who is on the autistic spectrum and has learning disabilities, because I was at home all day, he became more expectant of me spending time with him, so lost some of his independence.

On New Year's Day, 2017, I heard that my oldest friend, who I'd known since the start of primary school, had died suddenly.

At the beginning of February a former colleague, who knew I was knitter, sent me a link to a newspaper article about the search for knitters of ganseys. I've always been a knitter, my mother taught me the basics and then I taught myself other skills over the years. I did a lot of Intarsia work when my children were small, then more for nieces and nephews. Knitting was a way to be productive and creative when I was an RAF 'wife of' with young children. I'd seen ganseys exhibited in Whitby, but never thought I could make one. My daughter persuaded me to get in touch, so I did.

The first yarn arrived and I looked at the pattern, thinking I'd never manage, but I plucked up the courage to start and found myself hooked. It became my routine to do what I needed in the mornings, then take myself off into the spare room in the afternoons to work on the gansey. It was sometimes difficult, I made mistakes and had to backtrack, but the sense of achievement as I progressed was fantastic. When I had to wash the finished garment, I was terrified it would go wrong, but I managed not to shrink or stretch it and sent it off.

When I realised the second gansey, one I'd received part worked by a knitter in America, was going to be worn by one of the skippers, Liz, I was thrilled. Looking after the stall when Astrid and Janice were on the Williams II on her official return home after the Round Britain trip was icing on the cake.

It's been an honour and a privilege to be involved and I hope I can be a part of the project in the future. It's given me back a sense of purpose and a feeling of achievement. It just makes me proud to be part of something so worthwhile.

Wendy Hepple – Oxfordshire

It was an honour to have been selected to knit one of the Watch hats for the exciting Blyth Tall Ship Project. When I read about the Williams Gansey Project I knew that this was a way in which I could contribute to this worthwhile venture with its aim of bringing the largely forgotten achievement of the discovery of Antarctica back into the public consciousness in its bicentennial year. My husband is from the North East and most of his family members still live in the region (although I was born in the Midlands, and we have now lived in land-locked Oxfordshire for over 25 years). Over the generations there has been a strong family connection to the shipbuilding industry. I have been inspired by my visits to the area over the years, and by what I have learnt of its history and its relationship with the sea.

I chose to knit my Watch hat whilst our 18 year old daughter, who is taking a gap year before university, was volunteering in Madagascar. Her work there involved sailing around the islands (albeit in a rather different craft to the Williams II!), and diving to collect marine data for the Madagascar Research and Conservation Institute. The marvellous opportunity she had made me think of the different and valuable opportunities this project was providing for other young people – not only those who will crew the ship but also those who have learnt the engineering and traditional boat-building skills needed to restore the Williams II. I hope these experiences will stay with them as I'm certain my daughter's will with her.

The hat was a joy to knit – the wool beautiful to work with and the pattern, with its different elements, a reminder of the history behind these traditional garments and the meaning of the symbols they incorporate. I hope this hat gives the same pleasure to its wearer as its knitter had in creating it.

Wendy Heppell
Woodstock, Oxfordshire

Margaret Hill - Oxford

I am not from the north east. I was born in Oxford and lived there most of my life until 12 years ago when I moved to Romsey in Hampshire. I started knitting when I was six and now I am 74 so that's quite a long time! My main reason for joining your project was because I love knitting and hadn't done a gansey pattern before, but I also have great admiration for all those involved in the Blyth Tall Ship Williams adventure and will be watching the progress of the Williams II next year knowing 'my' hat is somewhere on board. Best wishes to the sailors and knitters and thank you to Astrid and Janice for inviting me to join in despite my southern roots.

Gill Humphreys
Appleton le Moors
North Yorkshire

Without wishing to sound too grand, I was drawn to the Blyth Tall Ship Williams Gansey project by a serious of overlaps with my own life.

Though born in Kent, my family have strong links with Yorkshire, having spent many happy summers with my grandparents in Catterick, which often included a trip to visit relatives in Newcastle-upon-Tyne; my Mum and Grandad's birthplace. On one wet visit, my Auntie Mary found a pair of yellow plastic knitting needles and navy blue yarn in her sideboard and taught me, a young seven year old, to knit, something I've been doing almost continuously for more than fifty years! Having spent many hours as my Mum's apprentice, patiently and sometimes impatiently, holding skeins of wool, it was interesting that for the gansey hat it was my 92 year old Dad who returned the favour.

I moved to Yorkshire permanently in 1992 and after fifteen years in a small market town relocated to a moorland village of no more than 68 dwellings. I was soon a member of the local history group and as such collaborated on three books about the village: Appleton le Moors and WW1; A Baker's Dozen (the history of 13 properties) and Luminaries of Appleton le Moors. Whilst the village is essentially a typical small moorland village, it was clear from some of the more substantial buildings that Appleton le Moors had once had a significant benefactor in Joseph Shepherd, ship owner and businessman.

Joseph Shepherd was the youngest son of John Shepherd and Susanna Scoresby and the young Joseph left Appleton le Moors to learn his seamanship from his cousin William Scoresby Jnr. Seafaring was very much in the Scoresby family's blood, William jnr's father, also William, having been apprenticed to a Quaker ship builder, Mr Chapman, at Whitby in 1779 and making his first voyage in 1780. William snr's whaling exploits began in 1785. Such was his success at catching whales – he became something of a local legend - that he was soon tempted away from Whitby to London. In 1799 Scoresby snr called at Whitby on his way to Greenland and his youngest son, William jnr, then aged 10, stowed away on board.

William snr, a significant seaman of his time, was credited with innovating a method to penetrate solid ice, which he called '*sallying*' – when the ship was ice bound, crew men would gather on one side of the vessel and run to the other side until the ship was freed. He also invented the crow's nest, not as some believed to avoid collision with other vessels but to spot whales. I hope that the crew of the Williams will not need to avail themselves of either the crow's nest or sallying on their 2019 voyage.

So when the opportunity to take part in the Blyth Tall Ship Williams Gansey Project fell into my lap, it felt like fate and I could not resist. Good luck!

Janet Jackson – Walkworth, Northumberland

Having found knitting therapeutic after my husband's death, I was lucky enough to be chosen to knit a gansey. This has proved absorbing, interesting, at times difficult & infuriating – but always rewarding. Thank you.

Janet Jackson

Avril Johnson – Cumbria

I was born and brought up not far from the Cumbrian coast and always enjoyed going down to the sea. When I saw the Williams Gansey Project it struck me as a brilliant way of combining my love of the sea and of sailing with my love of knitting. We live further inland now, so we've decorated our garden summerhouse 'jokingly' like a beach hut with model sailing ships, ships' lanterns, cute flying seagulls and so on. I like to sit in there and knit – and that's where my hat was made.

Ewa Johnson
Bedlington, Northumbrland

South-East Northumberland and the beautiful Northumbrian coast have been my adopted home for over 20 years. This region has a long and proud history of shipbuilding, fishing and seafaring with the sea providing not only sustenance, but also a sense of wonder and adventure. I was fascinated to learn about the Williams voyage and the project to recreate the ship and retrace the explorer's journey.

I have been knitting since I was a little girl, and I am proud to carry on the tradition of handknitted seaman's ganseys and hats, a beautiful part of our coastal heritage.

Maggie Johnson – Craster, Northumberland

I have been knitting and designing knitware for over 40 years as a hobby. My main interest is in stranded knitting and I led several workshops in Newcastle and Northumberland. Some of my work can be seen on my Facebook page under the name of Beeswing Knits.

After seeing a gansey exhibition several years ago in the Grace Darling Museum, Bamburgh, I became interested in the gansey patterns and designs. Following my retirement from teaching, I decided to have a go and have completed two ganseys, a navy one and, more recently, a grey one which I believe is for 'Sunday Best'. I enjoyed the challenge of designing and knitting these two jumpers but, I must admit, the navy one was hard on my eyes. They are a labour of love!

When friends told me about the Blyth Tall Ship Williams Gansey Project, I was excited to read all about it and am full of admiration both for you and all the knitters. However, having so recently completed the grey gansey, I was reluctant to knit another one particularly in navy. I was delighted to discover that you were offering knitters the chance to knit a hat instead - a perfect compromise! I have enjoyed knitting your pattern and love the Northumberland flag pattern you have incorporated.

I am thrilled to be included in this ambitious project and I will follow its development and the voyage with extra interest because of the small part I had in it. Thank you for your vision and hard work in coordinating this project.

Christine Jukes - Holyhead, Anglesey

Dear Gansey Project, here at last is the finished gansey - size 50" chest with and extra one and a half inches on the body and sleeves. I've taken some photos with my son wearing the gansey and some of me holding it. I shall email those to you. I can't wait to see the ship in full sail with its gansey clad crew.

Jackie Jurke – Alnwick, Northumberland

After spending a wonderful day in Newbiggin by the Sea in August 2016, watching the Tall Ships sailing from Blyth I spotted an article in the Northumberland Gazette requesting volunteer knitters for the Antarctic expedition. As a very keen knitter with over 40 years' experience I am always looking for new challenges. I have tried Icelandic, Aran and recently tackled Sanquhar gloves. I love learning new techniques and the idea of keeping old traditions alive. As I grew up in Yorkshire holidays were always spent on the East Coast and I was fascinated by the gansey tradition and the different patterns for each coastal village or even family. I particularly love the Blyth Tall Ships pattern with the sails, the staithes, ladders and the Northumberland flag. The pattern was well written and relatively easy to follow (I was pleased you changed the shoulder section) and, like most knitters I know would be, I was delighted to find there was very little sewing involved. I very much look forward to tackling more ganseys in the future but first I hope someone on the Tall Ships Project will enjoy wearing one and let me know how the project is going.

Susan Kay – Macclesfield, Cheshire

The Blyth Tall Ship Williams Gansey Project caught my attention but I didn't think I could cope with knitting a gansey so the watch hat seemed an Ideal way to be involved. My effort is not without its errors but it certainly stretched my skills in reading charts and knitting with such dark coloured wool. I would like to send best wishes to all those involved in the project.

Elizabeth Kerr – Chinley, Derbyshire

Please find enclosed my finished gansey. Thank you for the opportunity to participate in this wonderful project. I hope this gansey meets with your approval, I very much enjoyed making it. I have to admit I finished it months ago and then I wanted to show it to friends and family first before sending it to you. So that took a while to organise, I'm so sorry this is now very late! If I can be involved in any way with more knitting I would love to help. I promise not to take so long to return it to you next time!

I look forward to hearing news of the ship setting sail and the journey ahead!

Nia Kirwin – Bury, Lancashire (knitted a hat and a grey gansey)

I found knitting the gansey fascinating as I have never knitted anything so intricate or with this type of construction. I wanted to join the project as I like to knit with a purpose but also because I'm interested by exploration expeditions, especially Arctic/Antarctic ones, but would not like to be directly involved but this way I can join in, to a degree, but from the comfort of home.

Joyce Luby – Duns, Berwickshire

I am a member of the Scottish Women's Institute and when I got the email asking if I would like to take part in this adventure I jumped at the chance. The thought of an SWI member knitting a gansey for an Antarctic Expedition was so exciting. The SWI have projects going on all over the world and now at the ice caps too! I am very interested in natural history and enjoy watching TV programmes about this kind of thing. Hope my jumper keeps the crew member warm.

Pauline McAdam – Hawick, Scottish Borders

Sorry it has taken me so long to post this. I waited until the women in the knitting group had finished their hats as I still had some extra wool and two of us needed slightly more than 100g to complete the hat. However, everyone is finished. I'm sending the extra wool back as it may be useful elsewhere. Fingers crossed everything is ok. I have a photo I will email to you. Thank you for giving me the chance to be part of the project. We are talking of organising a trip to Blyth at some point but haven't got beyond the talking stage.

Helen McCree
North Shields, Tyne and Wear

I enjoyed doing the gansey because I feel it is so important to preserve traditional skills and show that they are still important. When I was asked by a friend to do my first gansey, I decided that because I had been knitting since I was four, I would give it a go, but it was a steep learning curve. Each gansey brings its own challenges and rewards and in this case being part of such an interesting project has been an added bonus. I look forward to watching as the project progresses. Good luck to everyone.

Margery Meakes
Conway, North Wales

As a child, my mother taught me how to knit, a skill that I have enjoyed and developed over the years and one that I find very satisfying and rewarding. Therefore, as an experienced knitter and having created three ganseys over the years, I thought the Blyth Tall Ship Project was both interesting and challenging - it captured the imagination! Also, I wanted to contribute to a small piece of history - and what could be better!

Thank you so much for enabling me to be part of your amazing project. I was brought up in Cullercoats, so the stories of 'ganseys' are part of my history. As a textile artist, I have always wanted to knit a gansey, so it was a great privilege to make my first one for you.
Please accept my apologies for taking such a long time to complete the work. Unfortunately, my carpel tunnel problems raised issues for me, but with the help of my GP.... I got there in the end!! Now it is finished, I'm waiting for an operating date.
I wish you and the whole project the very best and look forward to hearing how it progresses.

Lesley Miller – Almeria, Spain

It Started with a Stitch

101

Ann North – Penzance, Cornwall

Encouraged by my daughter Genevieve Jeffreys who had earlier knitted a gansey for the project, I was delighted to be included in knitting a hat for the project. I live in a coastal farmhouse overlooking the Celtic sea where many ships of all kinds have passed by over hundreds of years. I will watch out for the Williams II sailing by and if conditions are right may get a photograph.

Pam Norton – Hexham, Northumberland

When I first heard about this project I was very interested... Then I saw my friend's pattern for the gansey. I realised that would be too much of a challenge for me.
But a watch hat... ??
Now that seemed more achievable for my arthritic hands...
I have enjoyed the process of following intricate instructions and watching the pattern form - albeit slowly. So this is me looking quite pleased with the finished garment.
Below is a short explanation of why the project was special to me.

Continued...

Pam Norton Continued

FRANK PLATT, my dad, was born in 1905. He was a sea scout, I think in Blyth. Much to his parents and teacher's horror, he ran away to sea in 1922 (aged 17). From Blyth he sailed as a deck hand on a transporter all the way to Sunderland!

Five months later he joined the merchant navy, was assigned to ship in Blyth as an able seaman. There began his career which saw him rise through the ranks to become Commodore Captain of the John Hudson Steamship company, for many years sailing from Blyth to Dagenham on their fleet of colliers. As times changed the company expanded and he commanded their first motor vessel which bought raw sugar cane from the Dominican Republic to the UK.

In 1960, his war service and changes in shipping generally caused him to leave the merchant navy to take up a position advising on the development of the St Lawrence Seaway in Canada which allowed grain to be transported by ship through the Great Lakes.

He and mum retired to the North East of England where he was granted a couple of years to enjoy family life. He died on the 2nd of November 1969.

Hilary Paton – Newcastle upon Tyne

Hi, I felt motivated to knit a Blyth Tall Ship gansey watch hat as my son, when he was a teenager and venture scout, had taken part in one of the Tall Ships races. Subsequently he also volunteered on one of the Tall Ships owned by the Cirdian Trust, Queen Galadriel. His main experiences took place on the North Sea and he always said it was very cold out at sea. I hoped a warm woolly hat might help the crew taking part in this project. I have been knitting since I was a young child, mostly basic jumpers and scarves for family and friends. My daughter has also followed me in knitting and we knit at all opportunities when out and about.

Russell Pettit – Tiverton, Devon

I wanted to take part in the Williams Gansey Project for several reasons but the main one was to be part of something that was giving to others. I also wanted to represent the male knitters. I'm not sure how many there are connected with knitting for this project but I hope by me taking part and telling others about it can get a few more involved. I enjoyed knitting this pattern and as I did I thought of who might get the hat that I was knitting and where it would go on its travels. I would never have the chance to do something like this journey, but I can be part of it in spirit by knitting. I hope whoever gets it enjoys wearing it and that it keeps them nice and warm. Thank you for allowing me to be part of this wonderful project.

Kammya Prasad - Mumbai, India

It all started with a post in one of the Facebook groups about knitting a Gansey for an Antarctic Expedition. Curious, I contacted the Williams Gansey team and received a mail describing the project. It was exciting! Being a knitter with relatively little experience, I thought this project would be perfect to build my skill; apart from the feeling that even though I might never see that part of the world, some part of me, in this way, would. I agreed to take it on and by first week of July 2017 I received my parcel containing the necessary materials.

Once the hot and humid summer months of Mumbai ended, I began the knitting, around November. It appeared to be an easy knit. But as I reached the middle where the patterns start, reality struck me. It was going to be herculean!

I would knit miles of the pattern by day and frog double the rows I had knitted by night, much like Penelope and her burial shroud. It seemed like it would never end. But giving up is not me. Finally, in February, 2018 I knit the final row. I was thrilled to see my efforts take such a beautiful form!

Now, it feels like I have finished my first marathon, and as difficult as it was, I am ready to take on my next challenge.

Eileen Pritchett – Stockport, Greater Manchester
(written as a letter to their crew member)

Note from a gansey knitter to a gansey wearer

I have knitted a traditional gansey for you on your adventures to come (and one and a half hats). I have so enjoyed doing this.

I volunteered for this because I come from two very long lines of NE English and Scottish women who knitted to put food on the table, but also to take care of those they loved. I learned to knit when I was four, listened to my mam's tales of fishermen wearing ganseys with their initials knitted into the pattern, knitted through the unfashionable years, watched with amusement as knitting became 'A Craft', but also delighted that the skill and artistry of knitting was being rescued from obscurity. In our high-tech society we have gradually become separated from our craftsman roots. Not only does this expose us to total reliance on the thin veneer of our civilisation for our ability to survive, but it also deprives us of the great satisfaction, pride and good old mental harmony that such handiwork brings to all of us who practise skilled hand crafts and who know the value of it in our lives.

I also volunteered because, frankly, my budget doesn't stretch very often to knitting with beautiful wool and having the wool 'lent' to me as caretaker, to convert into your gansey, has been an added bonus for me, which I have thoroughly enjoyed.

I took it to Sweden on a long motorhome holiday, but mainly I knitted it while sitting on my sofa or out in our back garden, listening to birdsong and watching bees, chatting to my neighbours and husband, watching TV, talking handsfree on my smartphone to my daughter as she travels and studies. This garment therefore holds a lot of memories for me and will now hopefully hold new memories for my anonymous wearer. The rows flew by, but each stitch was made individually, one eye on the pattern and the other on a screen or a book, or a view or someone else's face. I've never "kept time" before while knitting, so it was a surprise to learn it took a total of 77 hours, and the hats nine and six hours. I've sewn in the labels, the messiest part as my hands don't hold a sewing needle very well now - please feel free to 'remove & improve' if felt necessary.

The hats were to use up the wool left, then of course there was not quite enough for another one after the first. Having been well taught by my 'make-do-and-mend' mam, naturally I improvised. The result is not as beautiful as the proper hat, but who knows, it might help out someone who has lost their proper hat, and just needs something over their ears to keep out an Antarctic wind.

Looking at the finished gansey (and proper hat) I'm chuffed with how it turned out and I hope the anonymous wearer will be as chuffed to wear it. May it bring you comfort when needed and please remember while you wear it how much pleasure it brought me in making it for you.

Jo Quigley – Weymouth, Dorsett
(Jo and her husband wrote a song about the gansey project)

My husband and I visited France many years ago and were lucky enough to see the building of a replica of the Hermione. I thought I would be bored but actually found it fascinating, and we spent several hours there watching the work going on. We always said we would like to revisit when they sailed her out of the dock but unfortunately we were in Thailand at the time so didn't see her sail!

When I saw your project on Facebook it caught my imagination. A I thought if I did it I would have the pattern to knit one for ourselves. B I liked the thought of receiving updates occasionally as to where the ship was and what was going on etc.

I knew it was going to be a huge challenge for me, actually much bigger than expected. I wasn't as good a knitter as I thought I was!

However, I am so glad I have done it, I am proud of my work and have got so many of our friends interested in what you are doing.

We would like to wish you the very best of luck and hope that she sets sail soon!! Do let us know when that is likely to happen and we will try to be there.

With all Best Wishes
Jo & Roger Quigley

107

Mary Richardson – East Lothian

I was born and brought up in Eyemouth, Berwickshire, a small fishing town. As a child I remember my granny knitting ganseys (or 'gainsays' as we pronounced them in Eyemouth). We come to Blyth and Whitley Bay occasionally with our motorhome and have watched the progress on the Williams II with interest. This was a lovely way to become involved with this worthwhile project. I attach a photo of myself and my HAT!

Lucy Robinson - North Shields, Tyne and Wear

I learned to knit as a kid, but nothing I made ever resembled what I meant to do! About two years ago I randomly decided to start again and - many unidentifiable things later - the pairs of booties actually started looking like booties and I expanded my horizons into more interesting things. After sharing a photo on Facebook of my first Aran jumper, a friend sent me a link to the gansey project. I thought it sounded really exciting and it was great that there was room for so many people to play a part. I didn't hesitate to apply and was really happy when the wool for the hat arrived. Knitting with navy wool on such tiny needles bought its challenges and having to take out 20 rows to correct a mistake I'd made in the double moss stitch in row 8 nearly doubled the time it took to make! But it was great watching the pattern develop and wondering what adventures might lie ahead for its wearer. I really look forward to hearing from them during the expedition and hope the hat enjoys its once in a lifetime experience! It's the envy of all my hats - they wish they could come along for the ride (and so do I)!

Susan Sabourn – Whitley Bay, Tyne and Wear

I'm a member of a knit and natter group at St John's Methodist Church, Whitley Bay and we have been involved in knitting Christmas angels, Easter chicks, poppies, premature baby clothes, prayer shawls, items for Operation Christmas Shoebox Appeal and Trauma Teddies for Northumbria Police over the last couple of years. I was taught to knit Aran jumpers by "Auntie Emily" a family friend when I was a little girl. Talking to Astrid about the gansey project, I was happy to hear that I could knit a hat on two needles as I have never knitted in the round before. I enjoyed visiting the Tall Ships event in Blyth and have watched them sailing from the Tyne in the past.

**Carol Ann Scorey
Berwick upon Tweed**

Thank you for allowing me to become involved with the Williams Gansey project.

I was pleased to be able to knit a gansey because I was struck by the boldness of the whole adventure.

Having visited the Antactic Peninsular on a cruise ship which resembled a floating hotel, I have nothing but admiration for the crew.

I wish them good luck and every success.

I knew nothing about the Williams, and Captain William Smith.

I hope that the project gives the story the exposure it deserves, as it is bound to reflect well on Blyth and the North East as a whole.

Eilidh Scott – Edinburgh, Scotland

> Delighted to say that I've finished the gansey and will be putting it in the post today. Here is a pic of me with it (I don't like getting my photo taken as you can tell!). I only lightly blocked it as I was nervous about spinning it so you might need to do that. Really hoping that I can come and visit the boat and exhibition at some point and looking forward to following the progress of the project. The random man is my husband modelling the finished gansey.
>
> I've sent another email with pictures of kids knitting a sleeve in one of my knitting classes.

Iris Scott – Blyth, Northumberland

It's been great to follow the Williams II journey around Britain's coast, and very good to know too that the gansey I knitted has been appreciated by an intrepid crew member.

I am sure that the crew members will have faced many challenges on their journey. We knitters too have faced challenges, sitting in comfort, knitting away on our sofas - especially with the shoulder panels!

Each knitter will have had their own reason for wanting to be part of this project. For me, from the first stitch, I was taken back to my childhood growing up on Coastguard Stations. Dad, Jim Sweeney, joined HM Coastguard in 1947, first at Filey, then Holy Island. Next to Scotland, Collieston, where I was born, and Peterhead. After a spell in Northern Ireland, he was stationed at Tynemouth where he remained there until retirement.

It was a unique experience having this background, although I did not appreciate it at the time. It was a safe environment and a happy community linked together by a common purpose. The coastguardsmen would have seen war service and if these men could help save lives at sea, then anything could get sorted.

Dad's job was 24/7, his work and our family life just merged into one. We were all involved. The old rule of children being seen but not heard had to be followed. If there was 'a job on' it was best to disappear!

There was always talk about rocket lines, hawsers, breeches buoys, what was happening along the coast...and the weather. We lived the weather and during stormy times a quiet expectancy would prevail, waiting for an emergency call. Dad's office was always attached to the house and as I got older, I would be roped in as tea maker and general gofer. I loved to accompany him on visits along the coast, especially at night when surprise visits had to be made to the men on watch.

As I knitted up my gansey I realised I knew so little of Dad's naval career. Like so many men of his generation he did not talk of his war time experiences. Both my parents died when I was 24, so there was no one to ask. With Dad's service record, some memorabilia and writing to the naval archives I have put together in a fashion, his 22 years of Royal Navy history.

Jim started his working life in 1925 as an apprentice fitter at Armstrong Whitworths armaments factory, Elswick, Newcastle on Tyne. Because of his small stature he was often sent along pipes to clean them out! He hated it so decided to join the army, but was rejected, once again because of his size. He signed up for the navy because they believed he would grow. He did!

He learnt to be a signaller and served on some iconic ships including Empress of India, Hood, Repulse, Renown, Niger, Broadwater, Halcyon, Victorious and HMS Rodney from 1943 to 1945. He travelled the world.

In 1937 he was seconded to an onshore base in Shanghai when the Japanese invaded China. British sailors fled and Dad managed to take some harrowing photos as they passed through the streets.

In June 1940 he was part of HMS Halcyon crew. This ship helped to rescue 2,271 service men from Dunkirk beaches. He was mentioned in despatches for his signalling work. In October 1941 he was with HMS Broadwater on convoy duties when it was torpedoed by U101 and sunk. There were 141 men on board, Dad was one of the lucky 85 survivors. He was later still to be part of convoy duties, including to the Arctic circle.

As next of kin I applied on his behalf for the Arctic Star, and he was awarded the medal. Dad used to talk about how cold it was chipping away at the ice on the ship and drinking lovely mugs of hot cocoa!

From April 1943 for two years he was part of HMS Rodney's crew, the ship with the big guns.

As part of the build-up for the Dunkirk invasion, he spent two months at HMS Mercury, a signalling school preparing for what was needed. He was one of the lucky ones, being in the right place on the right ship at the right time.

Dad was always knowledgeable and dedicated to his job, wise and very generous with his time. He would give talks on safety at sea, mentored students at the marine school in South Shields and even spent time with the boys at Wellesley Nautical School in Blyth. He was awarded the MBE when he retired in 1973, sadly passing away in 1976.

He called every jumper a gansey and I can truly say that his nautical DNA has been knitted into every stitch of mine.

Jenny Smith – Abingdon, Oxfordshire

A friend of mine, knowing I am an obsessive knitter, sent me a link to the project on facebook in its early days. Excitedly I applied to knit a navy gansey, and was disappointed not to be chosen. But I did make a navy hat! Thinking my link with the project was over, I was delighted to hear knitters were still needed for grey ganseys, so here is mine.

It has been fascinating to knit for such an unusual project, and everyone I know is now aware of it. I am not a sailor, and could never have been involved with the practical seamanship or boat restoration, but to be part of it with a completely different skill is great – what a wonderful idea! I shall follow the progress of the voyages with great interest.

Thank you!

Vera Smith
Jedburgh, Roxburghshire

I wanted to be involved in the project when I first heard about it. I thought the jumper would be too much of a knitting challenge so I was pleased when I found out that I could knit a hat. I have always had a kinship to all things shipping as I was brought up on the banks of the Tyne in Wallsend. I would often fall asleep to the sounds of men working away, building and repairing ships. The arc welding lit up the night sky and the 'clank, clank' of the shipwrights at work was very soothing. Coincidentally, I married my own Captain Elliot Smith, a merchant seaman, who has travelled the world but never made the journey to the Antarctic, unlike Captain William Smith. I do wish you all a safe and happy journey. Bon voyage.

Sue Sommerville – Roundwood, County Wicklow

I really enjoyed knitting the hat and being part of the project. I first heard about it when one of my friends in France was knitting a gansey. I am now following the Williams II as she circumnavigates Britain. Another curiosity is that Cork man, Edward Bransfield was master on the original ship and the people of Cork are erecting a memorial in his honour. I wonder if the Williams II project knows about this event. http://rememberingedwardbransfield.ie/
I'm a spinner more than a knitter as it takes so long to spin enough yarn for anything more than a scarf, a cowl or a hat. I own three llamas and four Birman cats, both of which provide me with lots of raw material to spin. I am also a sailor and a member of the Irish Cruising Yacht club. I think both projects relating to the Williams II are wonderful and I congratulate you in organising all the knitters, the wool, the patterns and everything. Well done!

Dawn Speight – Morpeth, Northumberland

I grew up in Blyth and lived there until I left for university when I was 18, so a project connected with my home town was always going to be appealing. My dad worked in Blyth shipyards up until they closed - had he been alive still, he would have been very interested in the Blyth Tall Ship Project.

I've never knitted a gansey until this one. I found it an interesting project to work on. I'd never made arm gussets before nor joined shoulders together quite like this, so I feel I've added to my skills as a knitter. This made up for the rather endless plain knitting in the round, but this was quite relaxing.

It has been enjoyable to be part of a wider project - I will follow the ship's voyage with interest.

Jean Stewart – Blyth, Northumberland.
(Jean knitted the largest size and a grey hat)

It was my husband, Rab who drew my attention to the New Post Leader article about looking for knitters and I thought I would give it a go as I am quite a keen knitter and this was a new project for me. I found it difficult at the beginning as I had never tried knitting with a circular needle. Once into it I was determined to get it done, regardless of how long it would take. Kay (Atkinson) was very helpful in showing me how the shoulders were done in the pattern. Nice to see the finished jumper.

Jane Straw – Derby

I can't remember where I first saw the request for knitters for the William gansey project but I know I applied as soon as I did. Initially it was for the challenge of knitting a traditional gansey but when I wasn't lucky enough to be asked to knit a blue gansey I started knitting for Flamborough Marine Company after seeing them on BBC's Countryfile.

When I was contacted and asked to knit a grey gansey I jumped at the chance. Not only would the grey yarn be easier to knit with in winter, I could also empathise with those crew more. Having lived all of my life in Derby, as far from the sea as you can get on this small island, I have no concept of how it might be to set sail and spend weeks at at sea.

The other draw for me was the historical aspect to the whole project. The fact that the original journey was so momentous and yet has had so little recognition over the years seems so wrong. I really hope that all funding will be raised and the people taking part in following the Williams in Williams II will take something from this re-creation.

I, in my small part, am happy to be sending this gansey to keep someone warm and dry in what could be horrible conditions. I hope the wearer will appreciate the hours spent knitting love & hugs into every stitch. You are so much braver than me!

Clare Szurek – Duravel, France

(written as a note for the crew member)

Sending you this Gansy, knitted in the heat of the sun and finished in the winter chill.

They say that with every piece knitted you leave a little of yourself, a hair, a thread from your clothes and the like. I am ginger so if you find copper it's mine. See I knitted precious metal in too!

Well we are Clare, Rik and Luc (12) living in Duravel in the Lot department (county) in the South West of France.

We also have Rolo, Baggins and Jessie the dogs, and 6 crazy chickens.

We have lived here for 4 years, having moved here to support Riks mum. His parents have lived here for 30 years, his dad passed just over 4 years ago.

We love Northumberland and visited last when Luc was a babe.

I am delighted to have been a part of this project and hope this Gansy helps keep you warm and safe. If I was younger this would have been an adventure I would have wanted to be a part of. My last physical mad scheme was Tough Mudder in 2013. My knees have begged for mercy ever since.

Enjoy this life experience it will give you memories to treasure for the rest of your life. Take care of yourself, there are good wishes and safe thoughts knitted into every row. I hope this gives you comfort if the going ever gets tough

A chance 'Gansey' encounter

In September 2019 we were staying with friends at their gite near Cahors in south west France. They introduced us to friends Bob and Debbie (ex pat brits) and in the course of conversation the Blyth Tall Ship project came up and amazingly Debbie had a friend called Clare Szurek, also living in France, who had knitted a gansey for the Gansey project. We arranged to meet up with her in a nearby village, Prayssac for a coffee the following day. There we chatted about the project and her experience of knitting the gansey, which she thoroughly enjoyed albeit a challenge at times!

It's a small world!!

Des O'Meara

Linda Talbot – Hexham, Northumberland

Date	Hours spent knitting	Comments

I began knitting the gansey in the Autumn of 2017. Not long after starting I received a diagnosis of cancer of the thyroid. I had surgery just after Christmas 2017 and was unwell such that I had a break from knitting for about 3-4 months. I had no concentration!

I then had another break in May when I had a dose of radioactive iodine. I was advised the gansey may soak up the radioactivity! When safe to do so I restarted and I finished knitting the gansey on 23rd June.

I apologise for not keeping a knitting diary. But the knitting of the gansey has been with me through a rollercoaster experience.

I have come out the other end cancer free and this gansey is a testament to that.

I wish the sailor who wears the gansey a safe voyage full of adventure and interest and wish him or her good health.

Linda Talbot.

Total Hours Time is immaterial ☺

Lotte Taylor – South Shields (Materialistics group)

When Materialistics of South Shields were approached to knit for the above project, I quite fancied knitting a gansey, but knew it would be beyond my capabilities so I was pleased to discover that it was possible to knit a hat instead. Knitting a hat could not be that difficult could it????

My hat includes a fair amount of 'blood, sweat and tears', but I got there in the end and am delighted with the finished article.

Furthermore, on a recent visit to Woodhorn (museum) I discovered that as a Dane, my hat will be worn by a crew member onboard an ex-Danish Boat out of Svendborg, previously called 'Habbet'. To say Habbet properly... just say 'Hope' quickly, and you get it right. My husband and I are coming up to Blyth after Easter and would very much like to see the boat, if at all possible. As it turns out our son, who works at 'Catapult' and had been at the inauguration of the project, follows the work on the boat from his office window (unbeknown to us).

Barbara Thompson
Shilbottle, Northumberland

Dear Astrid & Janice,

Enclosed completed gansey. I have enjoyed knitting it very much. We have e-mailed you a 'photo of my son and me, he's wearing the gansey!

I have knitted for very approx. 150/200 hrs. It was difficult to keep track as I had longish gaps between knitting bouts.

Best wishes for the completed project. Hope to hear more from you soon.

Best wishes
Barbara.

Jenny Thorpe – Burgess Hill, West Sussex

I wanted to knit this gansey in memory of my dad Sam Coxon who died in 1993 and was buried at sea off Blyth with thanks to members of Blyth boat club.

He taught me to knit when I was 4 yrs old, having been taught himself while in the Navy. They didn't have any child-sized needles, only long northern style needles so I was given two 6 inch nails to start with. I took to it straight away and knitted my first cardigan for myself when I was 7 yrs old — no television, x-box or anything in those days.

I am so proud and privileged to be included in this project. It hasn't been the easiest knit, mostly due to the dark colour and weight of the yarn. I am pleased with the result and look forward to seeing it being worn for its intended purpose.

Hours spent knitting — 147

Maryhelen Toal – Ontario, Canada

My gansey experience

This is a collection of photos of my great grandmother Elizabeth and my grandfather, Captain Hailey.

I've had a long time to think about what I should write about my experience knitting this gansey. It's about the kinship I feel with the countless women who began knitting ganseys in the late 14th century, over 600 years ago. I watched my own hands making the same stitches in the same way and thought of those women and of the seamen they loved. It gave me goosebumps every time I thought of them!

After more than 200 hours of knitting, it's finally finished!

I consider it a tribute to my seafaring ancestors who came from Yorkshire. My great grandmother, Elizabeth, who was lost in the sinking of the Empress of Ireland (1914) ...My grandfather, her son, who became captain of The Empress of Canada, a Canadian Pacific ship that sailed out of Vancouver...His son, Alfred, my uncle, who was a sea captain and harbour master at Nanaimo B.C. until he was lost at sea....and my father who sailed as first mate and winchman on ships along the coast of B.C. during his summer vacations from university.

They never quit! So when my fingers were sore, my hands cramped, my arms aching and my eyesight blurry, I refused to give up. I dedicate this work to all of them.

And to the sailor who wears this gansey.

Know that every stitch holds a thought. a wish. and a prayer for your safety on this long voyage. May you wear it in good health... And enjoy it!

From me, with love and respect.

Libby Tucker – Caerphilly, South Wales

I'm Libby and I live just outside Caerphilly in South Wales with my partner and cat. I have lots of hobbies including swimming, going to the gym, knitting and other crafting. I wanted to be involved in this project as I have always wanted to knit a gansey pattern and never quite had the guts to give it a try.

Nel Uytterhoevan - Belgium

After someone posted a call for knitters on the internet I searched for more information about the project. I immediately fell in love with the spirit of it.

It is important to preserve and to (re)discover old traditions and techniques. Moreover it connects people with all kinds of skills from different nations and social classes.

I hope that everyone is enjoying the project as I did with knitting the ganseys. It was really fine for this project! When I started knitting the sweater, my little niece (two and a half years old) asked what I was making. I told her it was for a fisherman or woman who's going to make a big voyage on a boat. And of course the other question 'why' came! I showed her the boat and told her that Antarctica is so cold that he or she needs a warm sweater! And it is a warm sweater, I've tested it in the first snow we had here! My niece liked the photoshoot in the nature reserve. My sister took the photos. Friends and family are also interested and regularly ask for feedback. For the first time I had the opportunity to knit with the real five-ply wool for this kind of gansey. I really wanted to try this wool and my wish came true. It is good, very sturdy wool. I now understand the name 'fisherman's iron' for it. I am convinced further ganseys will follow in this wool. Even if the pattern contains traditional motifs, it still feels contemporary. I was surprised that the gansey I knitted fitted really great for me. It will be used as a template for the next gansey. I hope the person who is going to wear 'my' sweater will feel that it is made with care and love, and that he/she will feel great satisfaction at the end of the journey. I wish everyone 'een behouden vaart' (Dutch expression for safe voyage).

Winifred Waite – Washington, Tyne and Wear

I got involved with the Williams Gansey Project through my brother Robert who told me about it. Both he and my dad were miners who were keen on going fishing in their coble after work. Dad had gone to sea on the fishing trawlers from North Shields when he was just 13, he joined the merchant navy and when war broke out transferred to the royal navy. Living in Seaham, fishing was popular as a way of relaxing after coming back from the pit. I suppose he missed the fresh air and freedom after being down a mine. It also provided us with extra food after the war when times were hard.

Robert was interested when he heard about the project and knew I would want to be involved as I have always been a keen knitter. My sister and I had learned to knit from a young age and enjoyed knitting Aran, lace and Fair Isle, well there was no telly in those days!

I had read about ganseys and how important they were not just to the fishermen but to the families after fishing tragedies as a means of identification, but I had never got around to knitting one. The gansey project offered both a challenge and an opportunity to help support a great local voyage recreating the original Williams voyage of 1819. Interestingly my husband was born in Seaton Sluice the birthplace of William Smith too. Hopefully this project will give William Smith greater recognition for his amazing discovery of Antarctica as well as showing the skills and dedication of everyone involved in the project.

Thanks to Astrid Adams and Janice Snowball for their dedication and talents creating the gansey pattern and getting the amazing amount of support worldwide to create such a successful project for the voyage. It's been hard work but a privilege to be involved.

Liz Wallace-Frances
Scarborough, Yorkshire

> Just a few words about me! Liz Wallace, 67 year old retired nurse, currently living in Scarborough where I intend staying forever. I joined the project as I'm a lifelong knitter with traditional gansey patterns and Aran patterns being my first love. When I'm not knitting or sewing, I enjoy pottery although I'm not very good at it. I'm also an RNLI volunteer on the Scarborough Visits Team and fundraising committee. It's the best and most rewarding thing I've done in my life. In true RNLI style, I'll just say "see you later" to whoever wears this hat!

Janet Walton – Sunderland, Tyne and Wear

> First of all, thanks for the opportunity to be involved, I'm sending the hat in to you in the morning post.
> I really am so pleased to have taken part in the project. I've always had links to the sea; my father served in the merchant navy all his working life as did my brother. I've always lived in Sunderland spending many hours at the coast. It's been very enjoyable knitting the hat and feel proud to have been engaged with this local project. My grandsons have shown a great interest in what I've been up to and it has been lovely sharing the story with them. The youngest, Matthew, has the hat on in our photo

Joan Wardle
Seaton Sluice, Northumberland

> It is hard to explain what attracted me to the call for knitters for the Gansey project, except it was simple really - I'm a reasonable knitter and enjoy a challenge! I've lived near the sea in South East Northumberland all my life, I have ancestors who were Master Mariners and worked out of Blyth, one side of my family have lived in Blyth for generations. Put all that together with my fascination of the lives of the local generations who have gone before us, then it was never in question that I would apply to be a Gansey knitter.
>
> I am so proud to have been selected and to have been on the Gansey journey with others and involved in the wider Blyth Tall Ship project. I have loved being a part of the Gansey Project and helping Astrid and Janice bring their vision to fruition. I've learnt a lot about William Smith and his amazing journey, how not to be put off because you've never done something before, and to have belief in your abilities. I've become a "Friend of Blyth Tall Ship" and have the greatest admiration for the work they do with young people.
>
> I've also now have the good fortune to have gained a friendship with two very special people in Astrid and Janice. Who would have thought that knitting would enrich your life in so many ways - thank you Gansey Project.

Marina Warner
Coventry

My name is Marina Warner and I live in Coventry with my husband, John. Our house is about as far away from the sea as it is possible to be in the UK, but we both love the sea and the coast and visit it whenever we are able to do so. Although not being sailors ourselves, we do love crossing the sea on ferries and boats of all sizes. Our 97-year-old uncle served in the royal navy, joining as a boy seaman, and often recounts his time on the training ship.

I have a very good friend who lives in Ponteland, not a great distance from Blyth, and when I visit her, we often make a "pilgrimage" to the sea together, so I feel a strong affinity with the Northumberland coast.

As a keen knitter who has visited quite a few coastal towns, the tradition of gansey knitting patterns has always been of great interest to me, although I have been a little too intimidated to attempt a whole jersey! The opportunity to knit a smaller project such as a hat for my first gansey knit was very welcome, although it did test my skills quite a lot.

The men and women who go to sea, in craft of whatever size and purpose, are keeping alive the great maritime tradition of our island nation, and I hope this hat will provide warmth and encouragement to the Blyth Tall Ship Project participant by whom it is received. Our very best wishes will go with you.

Judy Watson
Seaton Sluice, Northumberland

As an avid knitter since the age of seven, I was very keen to be involved with the challenge of knitting a unique gansey, especially as I live in Seaton Sluice, the birthplace of William Smith.

The opportunity to be part of this amazing project involving knitters from all parts of the globe was too good to miss and I've enjoyed every stitch...and "unstitch"! ...and to have your work appreciated by the sailors is a very special bonus. Many congratulations to Astrid and Janice for their vision for this project, and their tireless enthusiasm to make sure we as knitters were well informed and kept "in the loop".

My 70+ fingers are showing and feeling their age now, but what a way to finish a lifetime's hobby!

Miriam Werwick
Prestwich, Manchester

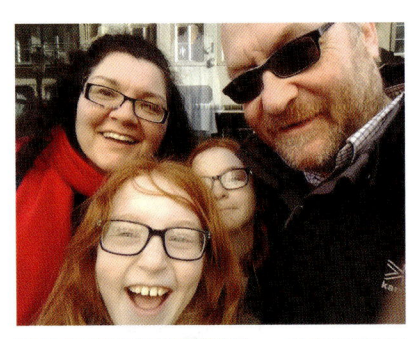

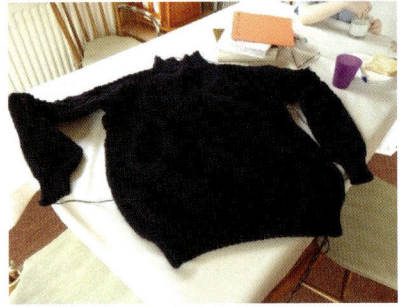

Please find enclosed by finished gansey. I can't believe it took me a year. It has a few cats hairs, I think, and a bit of chocolate at the bottom. I have kept it in my bag and have two young daughters, so had my gansey along with shopping, even washing powder. Apologies, think it might need a wash before wearing! Am enclosing the original pattern instructions, needles, and some remnants of wool, although found I just about had enough wool to finish the 42" size. My daughters Shira and Sammy are 10 and 9. Sammy was studying Antarctica at school last year, so thought it would be nice for them to have first-hand account of going to the place. Wishing whoever ends up with our gansey good luck and bon voyage. Thanks for putting together such an interesting pattern. I lent a lot along the way. Wishing all the team good luck. It sounds like an incredible adventure. Wish I could have joined you, but not to be at the moment.

Shaulaine White – Cambridge, Ontario, Canada

I've spent the last year knitting this gansey and it has accompanied me over hundreds of kilometres across Ontario. Parts of it were knit on the remote shores of Lake Superior, others along Lake Huron, but most of it was knit here in Cambridge, alongside many a pint! I asked to be included in this project as I have been fascinated by Antarctic voyages for many years. My husband's great uncle had been part of Sir Ernest Shackleton's crew, although no one is certain exactly which voyage as the records are not clear. Several years ago I bought my husband a copy of "South". I ended up reading it as well and found Shackleton's personal account so absorbing that I have been following Antarctic expeditions ever since, with the intent of one day getting there myself. I have endless respect for those brave enough to undertake these ventures to explore the remote or the unknown. I am honoured to contribute in any way I can and hope this sweater brings the wearer luck (and warmth!)

Janet Whitehead – Lowestoft, Suffolk

The project came along when I was busy with my Facebook page - The Lowestoft Gansey Project. It seemed an opportunity not to be missed.

**Bernadette Wild
Newbiggin by the Sea,
Northumberland**

Dear Astrid and Janice,

Please find enclosed a completed Watchhat.

As soon as I heard about the Project, I was very excited to join in. I have crocheted and knit for as long as I remember. I do more crochet these days (a bit easier on the arthritic thumbs) so I didn't think I could undertake knitting a Gansey. I was delighted to learn about the opportunity to knit a hat.

I have enjoyed making the hat very much and despite a few mistakes (corrected of course!) it was, dare I say, plain sailing!

I look forward to hearing more in due course.

Fiona Williams – Plymouth, Devon

Please find enclosed my completed watch hats for you.

One is the official navy version and one is my initial test one which I knitted in red. I am sure you will be able to find a use for the red one!

Why did I want to be involved in the project? Well, I always wanted to knit a gansey and the watch hats seemed like a gentle introduction. I love the combination of so many traditional crafts that the entire William Blyth Gansey project encompasses, the spinners of the wool, the knitters working on the ganseys and hats alongside the boat builders and sailing crews preparing the vessel for sea. All of us reaching back in time to connect with ancient skills to help push forward to future adventures.

The hats themselves come with some serious maritime credentials. I did most the knitting during my lunch breaks whilst working within the secure walls of Devonport Royal Dockyard. So the hats have 400 years of naval shipbuilding and know how knitted into their very fibres. The hats have also been worked on whilst I sat in our beach hut by the sea as storms raged outside. So the hats know about fierce seas and inclement conditions. Finally, the hats have been knitted by someone who has done a bit of tall ship sailing, so the very hands that held the knitting needles have also hauled their way up rigging, swept decks and set sails.

Thank you so much for giving me the opportunity to be part of such an interesting project.

Helen Williams
Stratford upon Avon, Warwickshire

This project caught my attention as all the knitting will be used by sailors, so it is needed. I found the history of the ship and the gansey tradition fascinating. Also, that tradition was being extended with a special pattern being made for the new ship's crew. I like a challenge and the pattern certainly provided that along with enjoyment and satisfaction at completing my hat! Last, but not least, my name is Helen Williams, so I feel proud to be part of the 'Williams Gansey Project'.

Elaine Wood
South Shields, Tyne and Wear (Materialistics Group)

Dear Astrid and Janice,

My gansey is finished! It's taken me much longer than I thought because in the early weeks I'd knitted for too many hours and ended up with a frozen shoulder which left me unable to knit at all for a couple of months. The shoulder is now much better.

I consider myself very lucky to have been chosen as a knitter, it's been a real pleasure to knit the gansey and to be part of such an exciting project. The Williams project is one of those things that captures the imagination and it has led me to many hours of internet research into William Smith, Tall ships, Olden days in Blyth, The South Shetland Islands and Antarctica.

The gansey wool was easy to knit with and of an even gauge throughout the cone. I'd certainly use it again if I ever knit another gansey. You might find that my gansey is firmer than the rest. I had to use size 2.25 mm needles to get the tension right, which of course makes it more closely knit. I hope it will keep the crew member extra warm.

I'm sorry the knitting is at an end. It has been great fun and has filled me with enthusiasm for the project. Everyone in the Tyne Valley must know about it by now! I'm following progress on the internet and I'll be there to see The Williams' set sail for the Antarctic.

With best wishes to you all

Elaine (Woods)

Chapter 8

Crew to Knitters Letters and Photos

Crew members' letters

To complete the circle, crew members were asked to write to their knitters to let them know a bit about themselves and their experiences on the Round Britain Trials.

Below is a copy of the letter given to all those on board the Williams II who received a Williams gansey and watch hat. Similar letters were given to those who were given their gansey and hat in person and those who received grey ganseys and hats.

The Blyth Tall Ship Williams Gansey Project

Dear

As a crewmember of the BTS Round Britain Trials, we are pleased to present you with your William Gansey and hat.

You will see from the label that your gansey was knitted by   from

Your hat was knitted by   from

The ganseys have been knitted by volunteer knitters from all over the world in a range of sizes. Our aim is to equip each crewmember of the Round Britain Trials and those who go on to sail on the Williams II in any future expeditions. The ganseys are expected to be a 'uniform' for the crews to wear on special occasions.

Because we have not been able to fit you personally, we are sending you a gansey and hat that we hope will fit you from the measurements you have provided. If you find that either the hat or the gansey do not fit you there should be some spare ones on the ship for you to try. If you do not keep the gansey or hat provided it is very important that you record on the sheets provided that you are taking a different one.

If you cannot find one to fit, email me and we will arrange for you to receive a gansey at a later date. Please take special note of the washing instructions. The gansey will not need washing during your trip - washing should always be kept to a minimum.

If you do want to wash it, it should be hand washed in warm water and spun on a spin programme in your washing machine. If you like you can add a spoonful of baby oil to the rinse water.
It should then be dried flat and allowed to air dry. Please never wash the gansey in hot water or dry it in a tumble drier.

The retail cost of the gansey is between £300 - £400 so please treat it with care. If looked after the gansey should last you a lifetime as a reminder of your involvement with the Blyth Tall Ship Project. We are asking you to send an email or two to your knitters via the email address at the top of this letter, letting them know which part of the voyage you have been on, how you became involved and what it has meant to you.

I will undertake to pass on your emails to your knitter. Photographs would also be very welcome. We hope you enjoy your gansey which was designed specifically for crewmembers of the BTS Project.

All best wishes for your trip,

Astrid Adams and Janice Snowball

Paul Allen – leg 7, 8 and 9   (a similar letter was sent to Lynda – his hat knitter)

Dear Janet,

I have returned home after a wonderful week on board Williams 2 and I am extremely pleased to now be the proud owner of the beautiful Gansey that you knitted. Thank you for this lovely jersey. It will always be treasured and well looked after.

I boarded the boat in Ullapool and we sailed as far as Peterhead, with a stop over in Scrabster and Wick. The Gansey was so comfortable and warm that it was worn every day of the trip. I know that it will now be a regular addition to my sailing kit when I sail on my own yacht, a 28 foot classic wooden boat. It will also be worn ashore and at home where I'm sure people will ask about it and I can tell them about the Gansey project and your involvement.

I have been involved with the Blyth Tall Ship since 2010 when I was employed through Port Training Services to set up the training workshop. At that time the workshop was an empty shed and so it has been lovely to see, over the years, how the training and facilities have grown. I left my role as trainer in 2015 and by that time we were delivering NVQ level 1, 2 and just starting the level 3. At the same time plans were being drawn up for the new workshop. Great oaks from little acorns do grow!

What especially pleases me is that I have helped set up a training centre that gives young adults the opportunity to learn new skills, get re-motivated and that builds self esteem and confidence. I know from my conversations with trainees that their time at Blyth Tall Ship is for many a life changing experience. It is also wonderful the way in which Blyth Tall Ship has become such a strong community project, that encourages the wider community to get involved in volunteering. The Gansey project is a perfect example of this and I am very grateful to you, Janet, for your involvement as a knitter.

I am also no stranger to Williams 2, or Haabet as she used to be called. I was a member of the team that went to Denmark to rig the boat and sail her back to the UK. I was also on the crew that sailed to Ayr when we went through the Caledonian canal and on the return passage I skippered the boat from Inverness back to Blyth.

As I wear the Gansey that you knitted, I will be reminded of the enjoyable times spent on board Williams 2 as well as it being a reminder of the happy and fulfilling years spent as a trainer in the workshop.

Thank you for your involvement and for the professional job you made when knitting the jersey.

With my grateful thanks and warm greetings

Paul Allen

Joe Boothby - leg 8

Joseph Boothby's message to gansey knitter Clare Szurek from Duravel, France.

Bonjour Clare!
My name is Joseph Boothby and I had the honour of wearing the gansey that you knitted for the Blyth Tall Ship Round Britain trip. Thank you! I have worked as an apprentice at Blyth Tall Ship Project for two years and have been employed as a carpenter for a further two years since then. When I first started at the project I couldn't sharpen a chisel, now I have rebuilt a huge amount of the Williams II: working on beams, planks and even a new rudder. I have also passed many craft skills on to others involved in the project. It was therefore a huge privilege after years of work to get to see my efforts being put to the test by the forces of the elements as the boat made its way around Britain.
I sailed from Oban to Ullapool, on the 3rd of May. The weather was much better than forecast though and it was cool enough that I wore my gansey the entire time! The views of this area of the west coast of Scotland were stunning — I have included some photos. I was quite nervous at first but the crew were great fun and we became friends and have all kept in touch since. Before we set off, the crew from an expensive yacht moored next to us told us their fridge was broken and we could have their food otherwise it would go off — so we ate like kings! Luckily, I wasn't sea sick but when I lay in my bunk at night listening to the wooden ship creak and groan as we sailed, I confess I was thinking of all the nails and wood I had fitted and hoping they would hold. Thankfully they did. Since coming back, I have joined a local sailing club and intend to have many more adventures at sea. I wanted to pass on my deep gratitude for the work and care you put in to making such a treasure of a garment that I will cherish forever.

Helen Boyle – leg 3

Well it is Sunday morning and there are a few engine issues. We are at Chatham historic docks in Kent some way up the river and it is pretty chilly but I am lovely and warm in my gansey and hat (thanks to you ladies) - see photo attached.

Tomorrow we plan to sail early out to open sea and over to Ramsgate. Will be going past the HMS Montgomery wreck that you may have seen on a recent FB video. It is loaded with live 60-year-old ammo. Eeeek.

Love from Helen x

Mike Bradburn – legs 3 and 4

Dear Genevieve and Sue

Thank you so much for the time and effort you put into the knitting. I am no expert at the technical aspects of knitting, but my wife was extremely impressed at the craftsmanship. I am looking forward to wearing them both, starting with two legs on the round Britain, then hopefully taking them up to 70 degrees North later on this year. I will be sailing from Chatham to Milford Haven, so I will be heading past your neck of the woods Genevieve. You should come and see us if we call into Penzance (which I think is very possible).

I got involved in the project purely by chance. Clive Gray came to talk at the yacht club and I was impressed with such a brilliant project. He mentioned that he was planning to have medical staff on board. This fitted in very well with my planned retirement so I went along for a chat. From that I got the job of organising all the medical cover through my contacts - but this was pretty easy and we have had a lot of junior medical staff who have signed up for legs of the journey.

I will either be on board as a sailor or a doctor, or probably a bit of both, but whatever, I am sure it will be a great trip, and I hope it will be life changing for the youngsters on board.

Best wishes

Chris Breeze – leg 1

Astrid, Chloe, Judy and Pam

Thanks again for the gansey and hat. Hats I can get that fit, but being 6' 7" tall, I have a real problem finding clothes that are the correct size. I often have to make do with the nearest thing. The gansey is great. I even have to turn the ends of the sleeves up, fantastic to keep your hands warm on a cold watch.

It also got the attention of people I met at Whitby, Grimsby and Lowestoft. They saw the gansey and asked if I was from the tall ship. Everyone was interested in the boat and the gansey.

A bit about myself :- I started sailing when I was 10, my father built a wooden mirror dinghy from a kit. In my late teens I started sailing on yachts whenever I got the chance. This was mostly on the East and South coasts as I lived in Cambridgeshire.

I skippered a racing yacht in an annual race for 16 years and then got into cruising. All together more relaxed!! I have owned three boats myself, apart from dinghies. First was a 17' trailer sailor then a 23' catamaran and now I have a 31' trimaran that I sail from the river at Amble.

Clearly these boats are very different from Williams II. I have however sailed on an Essex fishing smack out of Brightlingsea in Essex. Although they differ in size, the rig and sails are very similar, just smaller.

Ryan Brown – leg 5

Dear Jean and Pam

Thank you for taking the time to knit the wonderful gansey and hat. I sincerely appreciate the time you spent doing this and it was wonderful to have on the cold nights sailing. I took part in the Milford Haven to Holyhead part of the trip, sailing on the Irish Sea with the Bristol Port Company as an apprentice and thankfully had some very nice weather despite being initially delayed by the wind. Again, thank you so much for your wonderful gansey and hat.

Judith Brown – legs 8 and 9

> Dear Anne
>
> This is just a short note from me to say thank you for my gansey jumper. I spent the last two weeks on Williams II sailing from Oban to Peterhead via Mull, Mallaig, Ullapool, Scrabster and Wick and had a fantastic time. The weather was mostly fair although we did have a few rough times.
>
> I got involved with Blyth Tall Ship through my work with unemployed adults with a variety of mental health and other issues and have had a number of them attend the courses within the workshop to try to build their skills and confidence. Some have moved on into employment and all have improved their social skills and confidence.
>
> I used to work at sea many years ago and this trip was a chance to regain old skills and to meet more of the people involved with the project. I volunteered as the cook on my legs and thoroughly enjoyed cooking for everyone although as usual the cooker provided its own challenges. All good fun though and some great friendships made along the way which I am sure will last a lifetime.
>
> My gansey kept me very warm throughout and I really appreciate the time and effort as well as love that went into knitting it. I have attached a couple of photos which I hope you like from the trip.
>
> Perhaps I may meet you in person at the open day.
>
> (a similar letter was sent to Megan, her hat knitter)

Dave Brummit – legs 6 and 10 (two different people knitted this gansey)

Gill, Gill and Jen
Thank you very much for the hard work you put into providing me with my hat and gansey. I did two legs of the Round Britain trip, Anglesey to Ayr and Peterhead to Blyth. On both my legs my knitwear came in useful keeping me warm on night-watches. I broke 9,000 miles in my logbook on the Peterhead to Blyth leg – I sail a lot and will continue to wear my hat and gansey with pride in future races and expeditions.
Thanks again for your kind effort.

Alan Chapman – leg 1

Hi, thank you Pauline for knitting my gansey. It kept me warm through the long nights of sailing. At one point we reached 11.8 knots on the leg to Lowestoft and I finally got off the ship at Chatham after what seemed like a week of anchor watches and cold windy nights. Your gansey keeping me warm yet again. Thank you, Heather for my matching hat which kept my ears warm during the long cold nights.
Alan Chapman

Ian Cobb – leg 8

Hi Kay
My name is Ian Cobb and I sailed on the Williams II on the 8th leg from Oban to Ullerpool. I have retired from work and am a volunteer support worker with Headway Arts in Blyth who enable access to the arts through dance, theatre, drama and art to young adults with learning disabilities. Headway has a connection with the Blyth Tall Ship Project and it is through this connection that I became a crew member. The gansey fitted me perfectly and kept me warm throughout the journey. I, and most of the crew, wore our ganseys whenever we were in port and drew a lot of admirers and positive comments from people who saw us wearing them.
Attached is a photo of myself and my watch wearing our ganseys. I am the not so young chap at the front of the photo.
Thanks once again for such a wonderful gansey. I will wear it with fond memories of the leg and keep it hidden from my wife who has ambitions to wear it also!

Jen Davies – leg 9

Dear Kammya and Jane

I am writing to thank you for your wonderful hard work and talent in producing my gansey and hat for the Williams expedition. My name is Jennifer Davies, I am 46, a primary school teacher and mother to two fantastic teenagers. Ella is 15 and plays for our county netball team here in Durham. George is 18 and will soon start a degree in International Development in Norwich. For myself, I just completed my teacher training and have recently started my first role with five to six year olds, which is both challenging and rewarding!

Having always enjoyed outdoor adventures, I started sailing two years ago when I had the chance of a training day on board the Williams II, learning how to raise the sails, tack through the wind, do man overboard drills and drop and raise the anchor.

I was thrilled to be able to join the crew on its return into Blyth on the final leg of the Round Britain Voyage. It really was a highlight of the year and I was proud to be wearing your gansey and hat which not only looked smart and part of the crew uniform but was a remarkably practical piece of sea-going clothing that kept me warm, cut out the wind and repelled sea spray and rain in a way modern materials could never do. I hope to continue sailing with the Williams II and you can rest assured that the number one items in my kit bag will be my gansey and hat.

Clive Ducker – leg 8

Hi, my name is Clive Ducker.

I have been involved with the Tall Ship project for the past four years and five months, I first saw the ship in Svendborg, Denmark when it was being handed over to the project. I'm involved in training and assessing the candidates in the project from Level 1 to Level 3 NVQ. I was asked to do a leg and was chosen to do the Oban to Ullapool one.

It may have been the month of May, but it was very cold with a North wind while the ship was trying to make way in a northerly direction. So, the gansey and hat were much appreciated on the sessions spent doing night watches. The gansey became a near permanent fixture only to be removed when climbing into my sleeping bag.

I was able to pass on skills to the rest of the crew while away which gave a lot of satisfaction.

Chris Foster – leg 6

I joined the Williams II last weekend at Holyhead and left it yesterday at Ayr in Scotland after visiting Douglas, Isle of Man (leg 6).

The gansey I received had been knitted by Barbara and the hat by Lucy. Both were worn every day at some point and certainly served the purpose intended. They kept me warm when inactive and although I got very hot when working, they managed to allow me to cool down once I'd stopped working in a controlled way that prevented me from feeling cold.

Please pass my appreciation on to the two ladies who embraced me with their gansey on cold nights and mopped my brow with their hat during moments when sweating halyards.

I did fail to mention that I am a retired fire officer having served with Tyne & Wear Fire and Rescue Service for 32 years (half my life). Since retiring I have been active trekking the Pennine Way, Offas Dyke, Fjallraven Classic and trekking in the Himalayas to Everest Base Camp and Annapurna. I have also spent time on cycle holidays but have never sailed before - hence my interest in joining the Williams II expedition.

Alex Furnback – leg 8

Dear Sarah

I am writing to thank you for the gansey you knitted. It was a great and welcome addition to my sail from Oban to Ullapool. I had it on every day of my sail and can see me using it a lot more in my future travels. (a similar letter was sent to the hat knitter)

Dawn Furness – legs 1 and 2

Please forward my kindest thanks to Chris in Anglesey. It is the best jumper I have ever had. Superwarm and super cosy. I basically lived in it for the two weeks (and am still wearing it now)!

I feel so privileged to have been given such a beautiful gift. Huge appreciation to all the knitters!

Love and thanks,

Susan Gebbles – legs 2, 5, 6 and 10 (a similar letter was sent to the hat knitter)

Dear Margaret

Hello, I am the lucky person who has received the gansey you knitted; it is beautiful and so well put together, thank you so much for all your efforts. I particularly like the different elements in the design which reflect the Williams II project aims and Blyth's heritage. It is difficult to believe that it has no seams and has been knitted in one piece; I wish I had your talents! I am sailing on five legs: 1,2,5,6 and 10 as first mate. I can't quite believe that I have been asked to take part; it is a dream come true. As a teenager growing up in Liverpool I used to get all the Tall Ship prospectuses from the sail training agencies and hope that one day I would be able to participate. I was in my late 30s before I did finally learn to sail in a small two-man dinghy on Ullswater. I loved it so much I bought a small yacht, gave up my job as a marine biologist and became a sailing instructor. I still teach dinghy sailing in The Lakes during the main sailing season and work for an international boat delivery company the rest of the time. I am very excited to be taking part in The Round Britain expedition, a bit apprehensive about fulfilling all the duties that are expected of a first mate but it will be a learning experience for everyone. I am looking forward to meeting interesting people, exploring parts of the country by sea and working and living on a beautiful Tall Ship. The weather might well be cold and wet but I will have a wonderful gansey to keep me warm, thank you so much.

Clive Gray (CEO Blyth Tall Ship Project)

Dear Suzanne, Astrid and Jill

I am so delighted with my gansey and hat. I rarely take the gansey off (except when it is really hot) and have spent many days wearing it both on and off watch and in my bunk when getting up at a moments notice requires you to be on deck. They have kept the wind out and even the sea after a freak wave soaked me just as I poked my head on deck off the coast of East Anglia in a force 7 gale. I am Clive Gray, the Chief Exec of Blyth Tall Ship, Williams Expedition Leader and occasionally even the skipper of Williams II when cover is needed. The charity has taken up 10 years of my life so far and has used all my combined experience as a royal marine arctic and boats officer, global businessman and more recently, theologian. In truth the legacy of the project is less about a ship but more about the impact it has on the people who get involved in it. We see all sorts of responses, from disadvantaged and unemployed people gaining the qualifications and employability skills to get jobs, through to volunteers getting a sense of satisfaction, community and friendship that comes with supporting the charity. I hope that you have had enjoyment from following the expedition and knitting? I certainly can't quite believe how amazing the construction of this style of knitting is in keeping the harsh realities of the sea away from one's body and thank you again.

I completed five weeks of the Round Britain trip and I have attached a couple of photos of knitting in action! My gansey and hat will be used as a main part of my sailing gear as we go forward on to other expeditions.

All the very best

Clive

Nigel Gray – volunteer and rigging expert at BTS (grey gansey and hat)

This is just a quick note to convey my appreciation for the lovely "Williams" gansey that you made me. I feel so proud to wear it!! It's a perfect fit and beautifully made. The attention to detail is brilliant. It is very comfortable to wear too.

Today I visited the fishing heritage centre in Buckie. They were very impressed with the quality and design. There are quite a few examples of fishermen's ganseys in the centre. The hat is also comfortable and a good fit. As yet I've not washed it but will follow carefully the instructions you give.

Sue Hall – legs 9 and 10 – volunteer at BTS

Dear Marybelle and Janet

Thank you both so much for all the work and commitment you put into knitting my fabulous Blyth Tall Ship Project gansey and crew hat.

I sailed as watch leader on weeks 9 and 10 of the round Britain trip going from Ullapool round the top of the mainland and back home to Blyth. Both the gansey and the hat were absolutely brilliant, particularly during night watches in keeping me both warm and dry.

During my first week, the daytime temperatures were however very hot, so we were in shirts and tee-shirts (!) but in my second week my gansey and hat were back in daytime use too. I was particularly impressed by how well they kept off the spray, which often meant in the daytime I didn't need my waterproof top.

I immensely enjoyed my two weeks and while I have been involved with tall ships and classic vessels quite a bit one way or another, the challenge of working as a watch leader with young people with mental health issues was new to me. I think I rose to the challenge of mixing this with a varied group of crew of all ages. My 'day job' is as Volunteer co-ordinator at the Blyth Tall Ship workshop, where I work one and a half to two days a week as a volunteer myself. So I have known about the gansey project for some time now and never cease to be impressed by how Astrid and Janice's hard work has brought together knitters from all over the world. I have in fact knitted a crew hat myself, so have an idea of the time commitment made by the knitters.

So once again, thank you both so much for all your effort to knit my gansey and crew hat and for being part of this fabulous project

Claire Hughes – leg 10 – Berwick upon Tweed

Dear Carol

I was a member of the final leg crew of the Round Britain voyage having recently returned from rowing across the Atlantic. I had been missing the sea far too much so decided this was a great project to join and was excited to be using sails rather than a pair of oars!!!! It was also great to see the impact the project has been making and will continue to make in the wider community.

This voyage was significantly cooler than my Atlantic crossing so my gansey came in very useful - the fit was perfect! It was wonderfully warm and cosy and very much appreciated on late night and early morning watches in particular. We had a slightly murky start to our leg up in Peterhead, so again, it was fantastic having a cosy gansey to keep me warm rather than having to wear loads of layers of clothes. I can't even knit a scarf so how on earth you managed to knit the gansey, never mind the wonderful detail of the williams and the BTS logo on I will never know! Thank you so much once again for such a wonderful memento and I am looking forward to using it again this winter.

Roger Hardman – leg

21.4.2019

Dear Iris,

Thank you so much for the Gansey you knitted for the Blyth Tall Ship project. I picked this up a week ago when I boarded ship for my week crewing on the Milford Haven → Holyhead leg. For the first four days, when the weather was cold and the winds very strong, it was worn almost non stop. It is astonishingly warm and will be treated with care.

The trip round Britain is going well. We are sorting out all the little things (and a few big ones) that aren't working quite right on the boat. The leaks have been plugged, the engine is running smoothly and all the ropes are now doing what they should do when you pull on them. I hope to see you on the Williams II one of these days!

Roger Hardman

Inez Jacquemyn (Bruges, Belgium) – leg 10

Dear Beth and Susan

The last leg of the Williams II around Britain trip has come to an end a while ago and coming back to the hustle and bustle of daily work life, I forgot to thank you for knitting my gansey and watch hat, for which I apologise. I particularly appreciated the gansey and the hat during the many night watches on board of the last leg home; during a long watch from 3 to 6 am (sailing down the East Coast from Peterhead) and from 2 to 3:30 am (a magical and unforgettable anchor watch at Lindisfarne) the next day. The fit of the hat is perfect (even with my unruly curly hair) and kept me warm at all times. It's a very comforting and comfortable piece of kit. I crochet myself and am in awe and inspiration for the intricate pattern and the even stitches throughout.

Thank you again for the difference you have made to my Round Britain leg on the Williams II! I will be sailing the Norwegian coast this winter (on a modern ship) and will be all too happy wearing the hat and gansey on that (cold) occasion! I will be happy to spread the BTS message and explain the gansey project in particular during the voyage.

Dilwyn Jones – leg 9

Dear Emily

Just wanted to say thank you for all your time and effort in knitting the gansey for the Round Britain expedition last year on the Williams II, it was very much appreciated and came into its own on those long, cold night sails on the North Sea. The time spent on producing such an intricate, unique item of clothing makes the gift all the more special and your contribution makes an already memorable experience that little bit more special. Thanks again and keep up the great work.

Having a strong need for adventure and excitement combined with my love of the sea and all things associated with it, I have been involved with a number of different projects from a number of legs of the clipper challenge, several trips across the Atlantic in both directions on two other Tall Ships including a trip around the Caribbean islands and I was looking for my next adventure when the opportunity arose to get involved with the Blyth Tall Ship Project. Of course I jumped at the chance and loved every minute of it. Here's hoping that one day I get the chance to be involved in a trip to the poles, that would not only be amazing, I would get plenty of use out of my gansey too!

Cole Kelly – leg 1 (a similar letter was sent to Theresa who knitted her hat)

Dear Doreen

Thank you very much for the gansey that you knitted for the Williams II Round Britain Trials. I will cherish it forever and I don't think I took it off all week! I have been on a small sailboat once or twice but getting to train on a tall ship has been a dream of mine for most of my life. I was quite nervous at first, but I picked things up quickly I like to think. As you may know, we didn't get to leave when we were supposed to because of the dangerous winds but we got away at 5.30am on the Tuesday. Skipper Liz handed the helm over to me when we left the port and I got to steer the ship for over 30 minutes. So all the cool photos of us leaving as the sun was rising was me at the helm and not crashing into things! It definitely has to be one of the top 10 moments of my life.

We arrived in Whitby just around sunset and once we were secure, myself and a few others ran across Whitby and up the steps so we could get a better look at the Abbey. As an archaeologist I know quite a lot about the time period in which the abbey was founded so I gave a small lecture to my shipmates.

We had a sensible start the next day and headed off to Grimsby. It was a long tiring day and getting into Grimsby took ages. We sailed into a lock that filled with water and finally settled in about 9pm. We were quite tired and were not allowed off the ship until the next morning but a nice chap who has worked on re-planking the Williams II went out and bought us some beer!

We had most of the day off in Grimsby, had a shower and then a group of us went to Cleethorpes for fish and chips which was a nice day out. The weather was lovely so we set out at 5pm with a long night sail ahead of us. We took turns in watches of three hours on and three hours off and eventually got into our final port, Lowestoft.

A final dinner and then we had a nice night out. The next morning we made the ship spic and span for the next crew. It was sad leaving the Williams II but I've signed up for future voyages and will wear my gansey proudly till then.

Liz King - Skipper legs 1, 2, 3, 4, 7, 8, 9 and 10

Amanda - Virginia, USA, Claire – Gateshead, UK,
Ladies - Thank you so much for my wonderful gansey. I will wear it with pride.
As captain, my gansey will go through more wear and tear than most, but I have every confidence that it will stand up to the test. If it needs repairing I am sure the damage will be all my fault!
I have attached a picture of me wearing the gansey on board Williams II. Thanks to the generosity of our supporters and volunteers the ship is in great shape and we are keen to sail off on our spring expedition around Britain.

Claire MacKarill – leg 10

Dear Diane,
Thank you so much for taking the time to knit my lovely gansey. The skill that you obviously have is amazing. The gansey was handed to me as soon as I boarded Williams II. I tried it on and it fitted perfectly! I am not usually one to wear jumpers but there were only a few occasions during the trip that I took it off!.....so comfortable, pleasantly warm and very cosy (I think it became my new best friend!)
I was so lucky to be part of this adventure. As a young child I had a taste of sailing on family holidays, and loved it, but have never had the chance to sail since. Approaching 59 yrs old and going through a pretty stressful time caring for my elderly father etc, a friend mentioned about the project and thought I might be interested.
I jumped at the chance it was now or never! and knowing the trip was coming up really helped me get through the personal situation I was in the midst of.
The trip did not disappoint! Williams II is a beautifully refurbished ketch. Learning how this stunning vessel sails and being part of all that entails was such a treat. We had great weather, although a little more wind may have been good. The crew I sailed with were a great mixed bunch and I felt we very quickly made a good team. We came from very different backgrounds, different ages, and reasons for being on that trip. We laughed so much.
I was very happy to be part of that team and wore my gansey with pride. I think lifetime friendships have been made and such lovely memories. I know that each time I don my gansey I'll be reminded of my new friends and that once in a lifetime experience.

Sam Maddock – leg 4

I would like to send a big thank-you and an update on the fabulous Williams II gansey and hat, both of which I have worn almost continuously for the last week! I joined the boat at Weymouth for leg 4 of the Round Britain trip and was promptly issued with my new gansey and hat. This was a relief as the rest of the crew were already kitted out and once I got them on I was immediately one of the crew.

We had some technical difficulties with the boat and were unable to leave Weymouth until Tuesday lunchtime.

The trip was therefore going to be a long sail right around the South West – Dorset, Devon and Cornwall – then up into Milford Haven, South Wales.

We were divided into three watches and we were to sail three hours on, then six hours off for recovery over the coming days.

As we left Weymouth the whole crew were on deck, approaching Portland we hoisted the sails (quite a workout!), then it was our watch 6-9pm as it got dark and cold – this was definitely the time where our new knits were essential.

This first night was a challenge for all, we didn't get off watch until 10pm as we were needed to assist with a jibe (changing direction through the wind and moving sails). By this point the rain had set in and it was good to get below decks and into our bunks to get ready to be back on deck at 3am. However, at 1am the call of "all hands on deck" went up and again we had to assist the rest of the crew – we stayed on deck then until 6am before again falling into our bunks.

After this first night the weather settled and we understood the boat much more. We all had some time to relax in the sun and watch numerous dolphins playing off the bow.

In the early hours of Friday morning, we began our approach into Milford Haven, after a number of interesting hours navigating the busy port and lock, we tied up for the last time around 4.30am.

The experience has been one I will never forget. The gansey and hat will be items to be held onto and cherished – and used time and again on other sailing trips as they are clearly made to last.

I would like to thank the knitters of not only my own special hat and gansey but also everyone else involved. It is excellent personal kit but also makes all the crew look and feel part of the team.

Phil Manning – leg 8 (a similar letter was sent to the hat knitter)

I am more than pleased with my gansey, I'm delighted. I think the work that you have put in is exemplary and, although I don't normally wish my life away, I can hardly wait for the cooler weather so that I can wear my gansey out-of-doors. Alnwick does have a relaxed dress code but if I should have been seen outside in my new gansey, I think those nice men in white coats would think me in need of some care and attention. You may wish to know that I showed my gansey to a lady who is herself something of a needlewoman and she was very impressed with the skill manifest in the gansey.
Again, my thanks.

Ray McGinty – leg 7

Please pass on my thanks to Helen McCree for knitting my gansey. I took part in the leg from Ayr to Oban and was one of the watch leaders for that leg. I am hoping to be selected for the Arctic trip later this summer. The gansey is a beautiful garment and I was very impressed by the skill and care which has obviously gone in to making it. I'll always treasure it as a keepsake of the trip. I don't have an individual picture of me wearing the gansey, but here are a couple of the team photos where we are all wearing our 'uniform'. I am the old bloke with the beard in the back row of the picture third from the left.
Very many thanks indeed and I wish you success in any further projects you might be involved with.

John McMullen – leg 6

Dear Maryhelen and Margaret, I would like to say a big thank you for knitting my gansey and hat respectively.
They were both extremely smart and functional, especially going on watch at 3 am in a force 6 storm.
I am a 68 year old retired firefighter and had never sailed before. I live in a small, former mining village about four miles from Blyth so was keen to go on our local tall ship and experience "life at sea" and I thoroughly enjoyed it and am keen to have another go when I know my gansey and hat will be indispensable.
P.S. That's me operating Henry on our final clean of Williams II.

Kate Mojaat – BTS Supporter (working as coordinator, Port of Blyth)

(grey gansey and hat)

Dear Davina and Margaret
I am writing to express my deepest thanks for the Williams gansey and watch hat, which you were both so kind to volunteer your time and skills in knitting.
As a coordinator on the project for five years, I was pleasantly surprised to be included and thought of in such a lovely way.
Please accept my heartfelt thanks and best wishes for the future.

Connor Murphey – leg 5

Dear Jean and Ange,
Hi, my name is Connor Murphy, I'm 21 and am doing an apprenticeship with the Bristol Port Company and was given the brilliant opportunity to be a crew member for one week on the Williams II Blyth Tall Ship from Milford Haven to Holyhead. Little did we know that on arrival we would be given a personal gansey and that it just so happened to be knitted for over 200 hours by yourselves and I would just like to say how much they were appreciated - especially on the freezing cold night shifts but overall was an amazing all-round touch to the whole experience.
So I would like to thank you both so much for making an already amazing experience all the more special, and I really appreciate your amazing skill along with the love and passion you have for knitting the gansey.
Thank you ever so much.

Euan Montgomery
Skipper, legs 4, 5 and 6

I was the skipper from Weymouth to Milford to Douglas, Tobermory, Oban and Ullapool when I left the vessel.
My gansey got a battering but held up well.
Many thanks

Elaine O'Connor – leg 6

(a similar letter was sent to Tracy who knitted her hat)

Dear Sandra,

I would like to thank you for knitting my Blyth Tall Ship gansey which I wore for a leg of the Round Britain trip in May this year.

The trip was a fantastic experience for all involved and the group were all keen to wear their ganseys, not only for formal photographs and events, but also to keep them warm during the trip as the weather was generally rather chilly. I personally like to keep my gansey for 'best'. It really is a lovely item of clothing and I like to keep it for occasions in the autumn, winter and spring. It is great for days out when I don't need to wear a coat and also for outdoor activities which I do quite a lot of. I am keen to wear it on future ski trips to the Alps which I do regularly. It will be great both on and off the slopes.

Many thanks again for your work. It is much appreciated and the gansey is a real pleasure to wear.

Des O'Meara – leg 4 (a similar letter was sent to his hat knitter Sarah)

Hi Jennifer

A very big thank you for knitting my gansey. I know it will protect me from the wind and rain when I am sailing on the Williams II. My leg on board is from Weymouth to Milford Haven hopefully calling into the Scilly Isles and Falmouth and I expect to experience some foul weather during that week.

I think the fact that all of the crew on the Williams will be wearing these wonderful garments can only help us to feel part of the Williams' family. I'm sure everyone will be proud to wear them.

Blue skies

Des O'Meara

A proud gansey wearer!

David Patterson – leg 9

Hi Sarah

I should start off by telling you a bit about myself and how I ended up in the project. So, I'm David Patterson. I'm 26 years old and from the highlands of Scotland. So about three or four years ago I was working as a chef at a highland hotel when I got a bit bored one night and decided to go to Australia. A few weeks later I'm on a plane to Melbourne with literally my kilt a towel and one flip flop in my bag (it weighed 7.2kg when I checked in ha-ha) so anyway I spent a year working as a chef in a desert town called Marree in South Australia. I ended up coming home for my mother's wedding. When I headed back to Australia, I found myself without work. One day while job-hunting I spotted an advert for a ship's cook on a tall ship. Somehow I got the job and found myself the next day moving onboard a 1924 Baltic trader (her name is Southern Swan). Anyway, long story short, I ended up doing most of the maintenance work and stayed on her until my visa ran out. On the night of my leaving do the boss came to me and told me he would love to have me as an apprentice but because I wasn't an Australian citizen he couldn't do so, which made me realise I could probably keep doing this for a living and I should try to change from being a chef (never really enjoyed being a chef). Anyway, I head back home and end up working behind a bar in the same highland hotel as before, get chatting to people and get speaking to this woman who was a friend of Clive Gray. She gave me his number and I phoned him after my shift and asked if I could come and work and he said yes. So a few weeks later I moved down to Blyth where I volunteered for six months then I got onto the level 2 NVQ course where I worked as an apprentice for a further six months. Before my six months was up I managed to get a job at a traditional boat yard in Cornwall (Butler & Co Traditional Boats). This is where I am currently working as a shipwright, still in learning of course. On the expedition I was on the Ullapool to Peterhead part. It was a fantastic sail - had glorious sunshine for the whole week (which is rare for Scotland). I was the engineer/ watch leader on this trek. On one of my watches we managed to get her to 9.2 knots under sail which was fantastic sailing - well for me anyway. The rest of my watch were seasick as we had quite heavy seas. I love sailing in heavy seas so I was alright and on top of the world. That was definitely my highlight. I also did the pre-expedition sail training with Clive to train the doctors and such. I was also on board to deliver the Williams to London a few months ago. Honestly I couldn't be happier with the gansey. It's definitely my favourite hands down. I wear it at every opportunity. It also fits me perfectly which is great. Thank you Sarah for the gansey.

Thomas Price – leg 5

Pauline and Judy (hat and gansey knitters)

I am writing to you today to thank you for all your efforts in making my gansey and hat, for my recent trip aboard the Blyth Tall Ship Williams II. The whole week on board the ship I was kept wonderfully warm from head to toe thanks to the skill and expertise you both have used and shown in making me my gansey and hat. As far as the week went on board the ship, it was a very productive week from all 11 crew and we felt we learnt a lot from our week on board. We were given the opportunity through our apprenticeship at the Bristol Port Company that we are all currently undertaking at the moment. Although we were a little anxious about what to expect upon arrival in Milford Haven, we very much settled in quickly and got straight down to work and very much enjoyed our time on board the Williams II. Ultimately it was an opportunity that all of us would most probably never get to have again and we thoroughly enjoyed ourselves. However, I and my fellow crew disembarked at Milford Haven on Tuesday afternoon and we finally reached Holyhead Marina on Thursday evening after two long days at sea.

Once again I would like to take this opportunity to thank the both of you for all your efforts in designing and producing both my gansey and my hat. I will make sure I look after them and keep hold of them as a memory for many years to come. I hope that this letter to you comes as a nice surprise to you and you understand how thankful I am for all your efforts in ensuring that I was made to be a part of the crew.

Many thanks once again

Thomas Price (Apprentice Port Operative @ the Bristol Port Company).

Helen Quinn – Leg 6

Thank you very much for my lovely gansey and hat which kept me lovely and warm on my voyage. I was on Leg 6 from Anglesey to Ayr. We stopped at Douglas, Isle of Man and then sailed directly to Ayr for the best part of two days performing a watch system of three hours on duty and six off which involved sailing at night in the dark which was very different. We also had to endure a very bumpy Irish Sea which was a challenge to say the least but at least I discovered I didn't suffer from sea sickness ha ha!

I really don't know how I got involved as I had never sailed before and the only boats which I had been on were catamaran trips when on holiday and the odd ferry ride. But for some reason I was drawn into this project. Maybe it was because I was born and bred in Blyth and therefore grew up by the sea. Also I went to see the tall ships when they visited Blyth in 2016 and that's where I heard about the plans for Williams II. Then, purely by chance, I was visiting Blyth in March this year when the Williams II was due to set off on its 10-week voyage. I had the chance to take a look on board and learnt that there were places left to be part of the crew. I applied and was lucky enough to be accepted. It was a huge challenge for me, something completely out of my comfort zone, but it was something I felt I needed to do and was very grateful for the opportunity. I was made very welcome and everyone was very kind and patient. Overall it was an amazing experience.

Once again thanks for your amazing work and for being part of the Gansey Project.

Connor Redmond – leg 5

Thank you very much to Kay Atkinson and Alison Murry for taking the time to knit my Williams Gansey jumper, also to Jean Smith for knitting my hat for my leg of the Blyth Tall Ship round Britain trials. It came in very useful as a member of the crew on board the Williams II and kept me warm on the night watches as it got very cold. It's also a great memento to keep as a reminder of the trip that I can keep and use for many years to come. My name is Connor Redmond. The leg of the trip I took part in was from Milford Haven along the Irish sea to Holyhead in Anglesey. I was involved in the voyage through my work as a Port Apprentice at the Bristol Port Company as we have a close relationship with Port of Blyth. It was a once in a lifetime opportunity and although it was hard work, I enjoyed it a lot and felt I gained some vital experiences along the way. Once again, thank you for the gansey and hat, it was much appreciated.

Charlie Rowland – legs 1 and 2

Dear Jan and Sylvia

I wanted to thank you personally for the beautiful hat and gansey you have provided me with as part of the Williams Gansey Project. I have already had a number of compliments and lots of interest shown in the unique designs and have found them to be both a great look and very warm!

I have just finished an intensive three weeks of sailing and both items have served me very well. The first two weeks I sailed with the Williams II from Lowestoft to Weymouth and last week I was on a Tall Ship called the Tenacious, owned by the Jubilee Sailing Trust. They have become members of a carefully chosen number of personal favourites and I hope they will join me on a great number of other adventures. Navy blue is a good colour for me as I am a medical officer in the royal navy. I have a deployment coming up and I am looking forward to taking both items and seeing if I can get away with wearing them over my uniform – though I suspect I may not last long in them.

Thank you again for your skills time and efforts.

Judith Rust – leg 8

My main reason for embarking on a leg of the Round Britain voyage in May was to celebrate my retirement from teaching in December. I really wanted to do something different, exciting and have an adventure. The trip did not disappoint and I had an amazing week, meeting and working alongside a great crew sailing the Williams II.

I chose to do the leg from Oban to Ullapool. The scenery was amazing and the highlight was when some dolphins followed in the wake of the boat for a time.

We were only under sail a couple of times on the journey due to weather and tide conditions. But, to be honest, although it was wonderful to just be powered by the elements I enjoyed the whole experience of being at sea either under sail or engine!

The weather was, in the main, bright and fine but very cold. Knowing I was going to get my gansey and hat on arrival at the boat in Blyth I had only taken a body warmer and base layers so the gansey and hat were my constant apparel on the journey. It was really comforting just to pull it on over pyjamas too, when it was all hands on deck in the middle of the night. I am a jumper and wool fan so loved it from the moment I put it on. The gansey and hat proved their worth on a 36 hour non-stop sail when our watch was at the bow of the boat from 12 midnight until 3am. This was also the time that the weather had turned somewhat to a force 8, which is moderate to rough. After three hours on deck with the boat pitching every which way and spray and waves coming over the side the only bits of me that felt cold at 3am were my nose and toes. A true testament to the gansey and hat. This also made you think of sailors in the past and the conditions they had to endure.

I will always treasure my gansey and hat and it is a very special reminder of my wonderful trip with a great crew.

As a postscript I took part in an event called Joy, organised by Newcastle University, and I took my gansey and hat as a symbol of something that had brought me joy. Following on from this I have been selected to have a portrait taken and I am going to the Williams II this afternoon to be photographed on the boat (second from the right). I can only reiterate my heartfelt thanks for taking the time to be part of this project. I am looking forward to wearing the gansey and hat on a regular basis for my more land-based interest of fell walking. Bring on the cold weather!

Daniel Sadler - leg 3

I just wanted to write and ask you pass on my gratitude to Judy Watson, Jennie Ellis-Jenkins and Alison Dixon for making me my very special gansey and hat!

I am someone that never usually wears big "jumpers" ..opting more for layers because I find I get too hot. However, I absolutely love this gansey (and hat) both in design and function. Most of all I love it because it has been so skilfully hand made for me and fits (and works) like a glove.

I am lucky enough to have had a few tailored suits made for me over the years and I treasure the utter luxury of a skilled person making a personal garment. Having spent several years to obtain a degree in textile science I have some idea of the skill these three lovely ladies have employed to provide me much comfort and warmth on a 3 am night watch in force 7 winds as well as during the more placid times at sea and very much appreciate it. It is also a fantastic way to remember this voyage (and others) as well as the crew and ship itself. Wearing it makes me feel part of a special bunch of people.

Living in the of South East England I am not often in Blyth, but I hope to one day make your acquaintance and thank you all in person.

Kind regards

Michael Stanley – legs 1 and 2

Dear Janet and Barbara

Thank you for your time knitting the gansey and hat that I have pretty much been wearing non stop for the last two weeks. I was a watch leader on the Williams II and our journey was from Blyth to Chatham, a total of 369 nautical miles.

For the first week we travelled down the East Coast, first stopping at Whitby and then Grimsby. Whitby is a charming small town with a fishing harbour where we moored overnight at, looked down on by the ruins of Whitby Abbey – the inspiration for Bram Stoker's Dracula. Grimsby is at the opposite end of the spectrum. A big port showing the decay brought about by the slow decline of the British fishing industry. We moored at the Royal Dock and were treated to hot showers by the Grimsby and Cleethorpes Yacht Club. We finally arrived at Lowestoft where we travelled up the river and moored at Lowestoft Haven Marina.

After a crew change at Lowestoft we set off early and sailed to Harwich. This took us all day, giving the new crew the experience of raising and lowering sails, putting reefs in the sails and navigating through obstacles (sand banks, wind farms, fishing boats) that may prove to be one of the most challenging coastlines of the UK. We anchored in the River Stour which, due to problems with the lock at Ipswich, became our home for a few nights. This introduced the crew to 'anchor watches': two hours on and two hours off, taking turns to ensure the boat was held fast by the anchor. Our final stretch was through the busy waters of the Thames Estuary and the Medway to our destination mooring – a buoy outside the historic dockyard in Chatham. Thanks again for your time creating knitwear that became and continues to be our 'uniform'.

Kacper Stefaniak – leg 9

Hi Janice

I had the pleasure to be given a gansey that was knitted by you. I have to admit that since I got to know about around Britain expedition one of the things that I was the most excited about was ganseys... My dad is a keen sailor and I was aware of the practical values it has on the sea.

Your gansey was particularly helpful during night shifts. Having to wake up at 2am to sail three hours through the darkness, wind and cold was difficult at times, but the garment you knitted helped at least with my core body being warm! I also found it quite fascinating how universal it can be – during cold nights it was giving a lot warmth but when the sun was hitting it didn't make you feel like you're boiling. I've shown my gansey to the group of my friends and it was complimented a lot! The logo was knitted nicely, the collar is unusual (and hence impossible to get in standard clothing shops) and the colour goes well with many other clothes.

It's a piece that I'll be wearing not only during sailing trips but also onshore.

Tariq Tabiner – legs 8 and 9

Dear Gill and Richard

Thank you so much for knitting me my gansey and hat for my recent part in the Blyth Tall Ship journey around the UK. I was on board for two legs from Oban to Ullapool then Ullapool until Peterhead. It's been a fantastic experience and I was on board as ship's doctor but luckily no major medical events occurred. The gansey has been excellent and I've attached a pic with our watch in it above. I'm standing at the back. Thanks again for the effort and skill put into creating the gansey and hat, I love it!

Andre Touhey – leg 8

I have recently got back from Leg 8 of the around Britain voyage...wow what a great time I had!

Just wanted to fwd on this email to both my 'knitters' - Margaret East, who did a great job on my gansey, and Persephone Ditzel, who knitted my hat. I can't tell you how grateful I was to have such wonderfully warm garments while on the west coast of Scotland and it was great to 'look' the part too. I will treasure these and hope it becomes cold enough to wear them again soon, as I am from Sussex. Many, many thanks for the hours you have put into knitting these and for all the others. A truly a massive undertaking and a really nice project to sit adjacent to the Blyth Tall Ship Project.

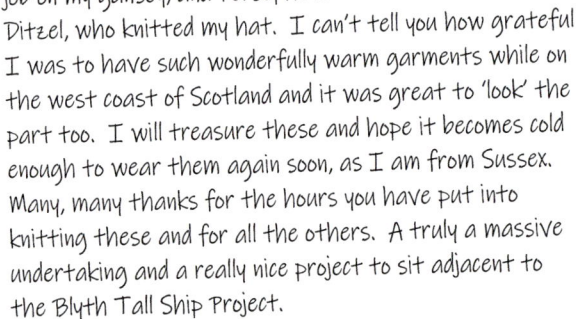

I hope to be able to return to the ship at some point when new trips and dates become available.

In the meantime, thanks again. Attached pic of us wearing them (I am second from the right at the front).

Sue Vicary – leg 8

I would like to thank both Anne for the gansey and Eileen for the hat. I was on the leg from Oban to Ullapool and it was very cold and rough at times so the warm clothes were really appreciated, especially during the night watches. I have attached a picture of my watch and I'm the person on the left of the picture. We had a little photo shoot when we got to Ullapool. I became involved with the project because my daughter lives in Blyth and sent me the link and I'd also done the Tall Ships Regatta from Blyth to Gothenburg previously. I love the tall ship experience although I always end up seasick, so probably ought to give up sailing as a bad job. The crew on my leg really jelled together as a team and we certainly want to keep in touch with each other and if possible meet up in Blyth when the Williams II returns in 10 days time.

Kathy and Lucy Ward – leg 4

My jumper was knitted by Sue and my hat by Eileen. My daughter Lucy received a jumper knitted by Doreen and her hat was made by Fiona.

Neither of us was aware of the Blyth Tall Ships project until just prior to joining our sail. We both have a love of Tall Ships but had never had the opportunity to sail on one, although we have been most envious of several friends who have. I remember sitting on the coast near Southsea watching the Tall Ships as part of the Cutty Sark Race when Lucy was small. As a family we have a very personal connection to Antarctica, as my great Grandfather was Edgar Evans. Edgar sailed with Captain Scott in 1912 and was one of the men who perished on the ill-fated effort to reach the South Pole. You will see then that the offer to participate united two interests. Finally, as we are both sailors (though had never sailed together before) it had to be fate that were offered this opportunity. When Euan mentioned the trip to Lucy, she was sitting on an aeroplane waiting to go to Budapest and so could not go. I asked if I could go instead. I had to request immediate holiday and pack a bag as I was to join the next day. Upon arrival in Weymouth, I learned that we had engine issues so would not be sailing as planned as the heat exchanger needed replacing and the spare would not be available until Monday. The engineer was a lifeboat man so fitting the part would depend on his call outs.

Kathy and Lucy Ward Continued

This meant that the crew would have time to bond, get acquainted with the ship and get much needed instructions as to how the sails were controlled. We had time to explore the many watering holes in port too, as can be seen from the image of crew plus ganseys in the pub!

It became apparent that we would not leave dock until Tuesday, by which time Lucy had returned from her weekend away. She was able to beg leave from her job to join us. She arrived late Monday, was put into a different watch team and given an adjacent bunk. So began our first time on board a sailing ship together.

Lucy likes to crew in yacht races and has sailed in the Caribbean, around Ireland and helped sail a boat on a delivery trip to Gran Canaria. She believes I am just a gin and tonic social sailor which explains why we had not sailed together before. It helped that neither of us had a clue how to sail a Tall Ship but were able to support each other and the rest of the crew in what was an exhausting, exhilarating trip of a lifetime. We had limited sleep due to the watch pattern, cooking duties and need to suddenly be "all hands on deck" if the sails needed resetting or when passing through the lock into Milford Haven harbour.

The gansey jumpers and hats kept us warm and cosy but were frequently hidden under waterproof oil skins and hoods as night temperatures fell along with copious rain. I had packed many layers but our ganseys were a useful extra and the love and time invested in them was very much appreciated.

I have worn my hat since on a trip up Snowdon which I climbed last week (first clear view from the summit and first without rain). Sadly, we do not have many photos from the actual sail but have attached what we did take where either a hat or jumper is visible. We did not want to risk a phone overboard and were also kept rather busy.

We were blessed with dolphins riding the bow wave on the Thursday when we could not see land on the leg between Lands End and Pembroke. This was a joyous event that drew all the crew on deck, regardless of whether they were meant to be eating or sleeping.

We were a totally new group of sailors, mixed in age and experience but who had to work as a team from the start to ensure we arrived safely and on time. We both thoroughly enjoyed the trip despite the physical effort required and had a lot of fun. Des, Chris and Billy kept up morale with sea shanties and tall tales. I am sorry that the ship will not get the environmental certificate needed to sail to Antarctica but feel privileged to have been able to crew on leg 4 and will never forget the experience.

Thanks again for organising the gansey jumpers; it's a very fitting tribute to the Tall Ship and fishing heritage of Blyth.

Acknowledgements

Firstly, and most importantly, I must mention Janice Snowball who shared my vision for the Blyth Tall Ship Williams Gansey Project as together we took a leap of faith into the unknown. Firm friends for many years, but with very different skills to bring to the project, we both contributed what we were best at. Janice's draughting skills brought the patterns to life and her close attention to detail meant that whatever she produced you could trust to be of the highest quality. We've come a long way since our first meeting with Clive Gray in February 2016, but it has been a road we have travelled together, providing mutual support when it was needed. I could not think of anyone better to work with.

My sister Claire Young has provided amazing help and support with the publication of this book and with the initial publicity for the project. I could not have done this without her.

To the knitters who tested early versions of the pattern and those who were prepared to finish ganseys that had been returned unfinished – thanks for the time and the skill you contributed to the project. You cannot know how much it was appreciated. This includes Doreen Armstrong (who very sadly has recently died), Sheena Cooper, Judy Watson, Claire Harrison, Kay Atkinson, Diana Blackburn, Mary Watkins, Caroline Taylor, Gill Charlton, Iris Scott and Eileen Pritchett.

Thanks to Sue Andrew who was responsible for raising the first £200 that got the Williams Gansey Project off the ground and to Blyth Masons for that important contribution.

Thanks to all the staff at Blyth Library who acted as a receiving point for returned ganseys and provided support during the project.

Northumberland County Council provided not only the funds to complete the project via their Community Chest, but also printed patterns and grids for us which saved a considerable amount of money and allowed us to knit many more ganseys.

Thanks to Jan and Russ Stanland at Frangipani who provided all the great quality 5-ply gansey wool for the project and also to Deb Gillanders (Propagansey) who pointed us in the right direction early on.

To our friends at Wonderwool Wales; we hope our very special relationship will continue once we are able to meet again.

Thanks to David Stocks and Gail Aldridge for proofreading – a tedious task! and to Mark Kelly for some of the photographs.

Special thanks to Jane Moffett who is publishing the book and to Anna Richards who typeset the pages. They helped to re-ignite a flame that had almost burnt out.

Also sincere thanks to Sue Hall for her steadfast belief in the book and her unstinting support.

And lastly, but by no meant least, a huge thank you to all the knitters who volunteered to get involved in the Blyth Tall Ship Williams Gansey Project. This book is a tribute to you and to the faith you placed in us to deliver what we had set out to do. There is not room to mention every knitter in this book, but even if you are not mentioned by name, your contribution to the project was still important and valued. I know there must be some disappointment that the original plan for the *Williams II* to sail to Antarctica had to be abandoned. This decision was not taken lightly but was unavoidable at the time. What we can say is that the Williams Gansey Project delivered exactly what it set out to achieve – and more. I think the selection of letters from crew members on the Round Britain Trials demonstrates how much the work that went into producing the ganseys and hats was very much appreciated.

Profits from the sales of this book will be donated to the Blyth Tall Ship Project.

For further information about the Blyth Tall Ship Project visit: https://www.blythtallship.co.uk

If you would like to know about sailing on the Williams II or holding an event on board, visit: https://www.williams-sailing.co.uk

You can read current updates on or about the project at the Blyth Tall Ship Facebook page.

There is a Friends of Blyth Tall Ship group. Visit Friends of Blyth Tall Ship Facebook page.

The Williams Gansey Project has its own Facebook page. Visit Blyth Tall Ship Williams Gansey Project.

To contact the author with any queries email – williamsgansey@gmail.com

Knitting patterns available
Gansey and hat - £15 + p&p
Hat, scarf and fingerless gloves - £6 + p&p
Sock pattern - £3 + p&p
(Available by post or electronically as a PDF file on request)

In memory of Doreen Armstrong